Process Development Division

AA003280

Presentations at the 2009 AIChE Spring National Meeting

Tampa, Florida
26 – 30 April 2009

ISBN: 978-1-61567-268-4

Printed from e-media with permission by:

Curran Associates, Inc.
57 Morehouse Lane
Red Hook, NY 12571

Some format issues inherent in the e-media version may also appear in this print version.

Printed by Curran Associates, Inc. (2009)

AIChE
3 Park Avenue
New York, NY 10016-5991

Phone: (203) 702-7660
Fax: (203) 775-5177

www.aiche.org

Additional copies of this publication are available from:

Curran Associates, Inc.
57 Morehouse Lane
Red Hook, NY 12571 USA
Phone: 845-758-0400
Fax: 845-758-2634
Email: curran@proceedings.com
Web: www.proceedings.com

TABLE OF CONTENTS

The Evolution of Process Control - From Hand-Turned Valves to Wireless Networks

Pete Sharpe, Greg Martin
Advanced Applied Technologies Group
Emerson Process Management
Austin, Texas

Abstract

Over the past 50 years, fundamental advancements in process control technologies have driven dramatic improvements in plants' operating efficiencies, productivity and asset utilization. The move from pneumatic to electronic controls improved response times and control loop performance. Distributed Control Systems (DCS) brought another step change in capability and established a platform to build Advanced Process Control (APC) applications. Fieldbus technologies and intelligent field devices are driving another transformation in the process industries with advanced diagnostics and asset management functions. Wireless control networks are poised to be the next major disruptive technology that could proliferate cheap measurement devices into services that have previously been cost prohibitive. These technology shifts have also changed how control systems are designed, installed and maintained.

This paper will take a historical perspective on how instrumentation and process control has evolved over the years and what each major innovation has meant to the process industries. A review of new process control technologies in their infancy will provide a vision of what the future might bring.

A Guide for Signal Processing of Sensors and Transducers

Himanshu Dehra
American Institute-Industry Forum on Energy
1-140 Avenue Windsor, Lachine, Québec, H8R 1P7 Canada

Introduction

The gastrointestinal tract, the lungs, the kidneys, and skin make exchange of materials and energy between the internal and the external environments. A steady state is achieved by regulatory mechanisms involving the balance between the inflow and outflow of the internal environment that stabilizes the composition of the internal environment [1]. ***The tendency to regulate the internal environment so that it is maintained in a steady state is called homeostasis.*** Our bodies have feedback systems that regulate our internal environment. The feedback systems make use of storage depots and numerous feedback loops. The monitoring of plasma calcium is a good example of negative feedback. The bones constitute large storage depots for calcium, for the plasma to withdraw these storage supplies in times of need. Our body's homeostatic regulatory systems are represented by feedback loops. The feedback is considered negative, when it is compensating or negates any change. The negative feedback is essential to stabilize a system.

The objective of the paper is to present an analogy for signal processing of our sensory organs, our communications through speech, our written words, and our associated pictures. The circuit analysis and synthesis for sensory organs of a human body is achieved by analog mixed-signal circuit design of a parallel plate photovoltaic amplifier [3, 6]. The logic NOT gate for a parallel plate photovoltaic amplifier is a source, sink matrix inverse constituting noise filters that offers immunity to the internal environment. A parameter that is closely related to the logic 0 and logic 1 levels is the noise margin.

Noise Filters

The noise filters are classified as per source signal of unwanted frequencies from solar power, electric power, light power, sound power, heat power, fluid power and fire power [5]. An acoustic filter is an electrical analog circuit of various combinations of RC feedback circuit with an operational amplifier and is used to filter unwanted frequencies of oscillations from a power system [4]. It is a network with selective transmission for currents from a power system of varying frequency. The unwanted frequencies generated from a power system are removed by using an operational amplifier with different combination of filter arrangements. An operational amplifier is an integrated circuit that consists of several bipolar transistors, resistors, diodes, and capacitors, interconnected so that amplification can be achieved over a wide range of frequencies [4].

The action of filtering the frequency from a power system is based on the variation in the reactance of an inductance or a capacitance with change in frequency [4]. The band of frequencies that can be removed from a power system can be at the low frequency end of frequency spectrum, at the high frequency end, at both ends, or in the middle of the spectrum. The filters to perform each of these operations are known respectively as low-pass filters, high-pass filters, band-pass filters and band-stop filters [4]. There are many configurations of design of filters. The filters are divided into passive and active configurations. The passive filters are less effective simple circuits constructed with resistors, capacitors, and inductors. The active filters are useful in providing an effective filtering action than passive filters. The active filters require a source of operating power.

The criteria for definitions of filters for noise filtering is based on areas of energy stored in a wave due to noise interference, speed of wave and difference of power between two intensities of wave [5, 6]. The noise source signals that can be filtered are considered from systems of solar power, electric power, light power, sound power, heat power, fluid power and fire power [5]. The closed loop negative feedback of an operational amplifier is obtained by connecting the output to the inverting input through a potentiometer. The negative feedback is useful for volume control of a highly versatile amplifier [2, 3]. The gain in closed loop configuration is directly proportional to the feedback loop resistance. As an illustration of negative feedback configuration for signal processing and filtering various sources of noise, a photovoltaic amplifier was constructed with a pair of glass coated photovoltaic modules forming a parallel plate channel with a plywood board and connected to a potentiometer of variable resistance up to 50 Ω [2]. A potentiometer was a wire-wound circular coil with a sliding knob contact. The schematic of the experimental apparatus for a parallel plate photovoltaic amplifier connected to a potentiometer is illustrated in Figures 1 and 2.

Experiments

As illustrated in Figure 2, the pair of photovoltaic (PV) modules used was connected in series an amplifier circuit for generation of electric power with a potentiometer of varying resistance up to 50 Ω [2]. The electric power generation from two PV modules was not at maximum power point because of resistance of potentiometer across the circuit [2]. T-type thermocouples were used for collecting thermal measurements from photovoltaic amplifier. Three thermocouple sensors placed at the top, middle and bottom locations in PV module, air-passage and back panel were used to measure their local temperatures at the defined positions. Two thermocouples were used to measure the inside test room air temperature and ambient air temperature. One differential thermocouple measures difference in temperature of the air passage between top and bottom sections of a photovoltaic amplifier. The copper-constantan junction (T-type) thermocouples used for measurements generate voltage signals to the data logger. All the thermocouples connected to the data acquisition system were machine cut and made from single roll of wire. The thermocouples were calibrated using heat flow meter (model: Dynatech R-Matic). In order to calibrate the thermocouple, one differential thermocouple was made from the roll of thermocouple wires and was fixed on the wooden test sample (with known RSI value) which was placed inside the calibrated heat flow meter to generate the voltage output signal. The signal generated was compared with the output given by the heat flow meter. Based on the measurements, uncertainty in thermocouple measurements was estimated at ± 0.05 °C [2]. The thermocouples exposed either to solar or surface radiations were shielded by cones made of tape to avoid radiation effect in temperature measurements [2].

The hybrid airflow created through a parallel plate photovoltaic amplifier was by natural wind, or the buoyancy effect in the absence of wind. The fan pressure was used to achieve higher air velocities by operation of the exhaust fan fixed on opposite wall with respect to the test section wall. The slight negative pressure was induced for drawing low air velocities in absence of wind-induced pressure from the inlet damper into the collector through the test room. As is illustrated in Figure 1, only dampers D1 and D2 were used for the present work for creating hybrid airflow with or without use of fan-induced pressure. The manually operated damper at top of the test section placed inside the outdoor test room was used for receiving pre-conditioned fresh air into the room. The exhaust ventilation fan was rated at 270 CFM (ft^3/min) (0.127 m^3 s^{-1}) at standard atmospheric pressure so as to create between 10-25 Pa of negative pressure depending on the air leakage from the outdoor test room [2].

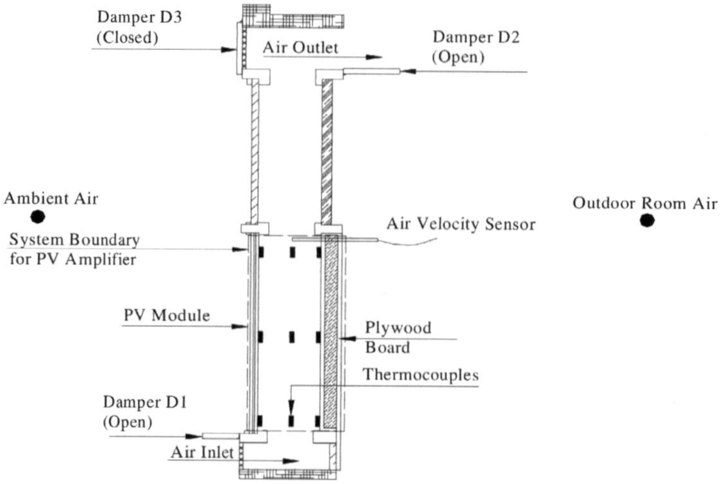

Figure 1. Schematic of a photovoltaic amplifier with location of sensors

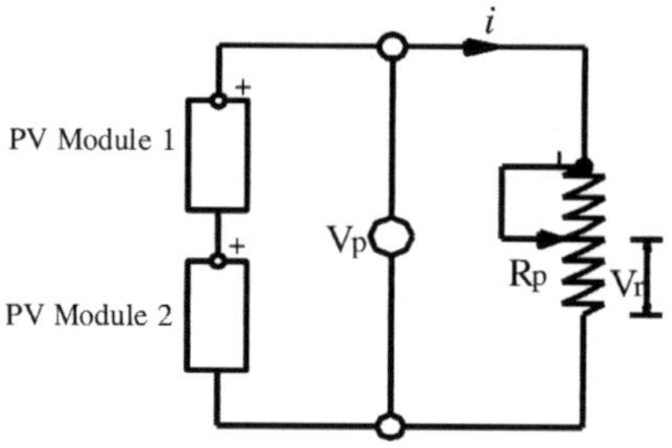

Figure 2. Electrical circuit diagram

The three dimensional effect was reduced while operating fan-induced hybrid airflow by developing air velocities by the induced draft fan fixed on the opposite wall so that it was placed at a distance from the ventilated wall of test section containing photovoltaic amplifier. It was validated by one of the experiments that measured surface temperatures of the two adjacent PV modules were within 0.5 °C under similar operating conditions [2]. The thermocouple outputs, current, voltage, solar irradiation and air velocity signals were connected to a data logger and a computer for data storage. The air velocity sensor was placed perpendicular to the walls of the collector to record axial air velocities of the airflow near the outlet of the parallel plate air passage [2].

The operation of air velocity sensor (make-Dwyer instruments, model no. 640-0) was based on the current output and has four field selectable ranges for air velocity measurement from 0 to 60 m/s and has accuracy of ± 5% for range of 0-1 m/s at the operating air temperature range of 0-100°C [2]. In order to minimize boundary layer effect in air velocity measurement, air velocity sensor was placed at middle air-gap width at a distance of 45 mm from either of the walls of the collector. To minimize turbulence effects in air velocity measurement at the entrance and exit region of the collector, velocity sensor was placed at outlet (at 1100 mm from the bottom) of a photovoltaic amplifier with total air passage length of 2100 mm in the test section [2].

The solar intensity measurements were performed with the photometer sensor fixed on the exposed wall fixed with a photovoltaic amplifier [2]. The calibration of the photometer was performed with the calibrated pyranometer of known correction factor [2]. The output signals from the photometer and pyranometer were compared (at known solar intensity) with the multi-meter and correction factor for the photometer was obtained [2]. The measurement sensor in the photometer was based on a silicon solar cell. It was assumed that the radiation characteristics of a photovoltaic amplifier and photometer sensor were in the same radiation spectrum thus minimizing the measurement error induced due to incident angle dependency of surface radiation properties. Other reasons for using the photometer were its light weight, easiness and robustness in fixing on the wall in comparison with the similar experimental setup performed with the heavy as well as delicate pyranometer [2].

The voltage and current outputs from the transducers placed on different positions on the equipment were connected to the Agilent 34970A data acquisition system [2]. VEE Pro software is used to design and build control functions for collection and storage of the experimental data [2]. The collected data in the data logger was imported to the computer in Notepad and Microsoft Excel programs with use of VEE Pro software. The Agilent 34970A data acquisition system was calibrated with the multi-meter by comparison of output signals from the sample measurements. The measurements collected in the data logger were up to six significant figures [2].

A Matrix Inverse Model for Noise Filtering

The assumptions applicable to the simulation model are [2]: (i) quasi steady state heat transfer analysis has been performed assuming a vertical channel for PV module test section; (ii) uniform average air velocity distribution; (iii) temperature variation only in y-direction (vertical), being taken as lumped in other directions (x-axis and z-axis); (iv) air properties are evaluated at film temperature of 300 K; (v) negligible heat transfer from side walls/insulation panel and room air zone; (vi) conduction (diffusion) equation for performing energy balance on air nodes is not taken into consideration; (vii) negligible thermal storage capacity of PV module test section; (viii) no infiltration or air leakage sources from the test section; and (ix) ambient air and room air temperatures are specified.

The procedure for numerical solution involves [2]: (i) construction and placement of nodes; (ii) energy balance on the nodes; and (iii) solution of system of algebraic equations. The PV module with nodes 0 to 9 forms passage for solar air ventilation with the insulation panel from nodes 20 to 29 [2]. Nodes 10 to 19 are for air flowing through the photovoltaic solar wall section [2]. The conjugate heat exchange analysis is performed for one PV module and corresponding areas of air passage and insulation panel in a photovoltaic amplifier. The grid size is 3 X 10, with grid spacing of 0.09 m in y-direction and 0.045 m in x-direction [2].

The finite difference approximations are used to develop algebraic nodal equations for each node in the computational grid. The nodal equations are developed by performing an energy balance on surface and air nodes in the control volume i.e. control volume finite difference method (CVFDM) is used for representing algebraic nodal equations [2]. The general energy balance equation for 'i' surface nodes of PV module is written as:

$$\Delta Q(amb)_{i,o} + \Delta Q(cond)_{i,i-1} + q\,\Delta V + \Delta Q(conv)_{i,k} + \Delta Q(cond)_{i,i+1} + \sum_{j} Q(rad)_{i,j} = 0 \quad (1)$$

Where, $\Delta Q(amb)$, $\Delta Q(cond)$, $\Delta Q(conv)$ and $\Delta Q(rad)$ are heat exchange terms due to ambient heat loss, conduction (in vertical y-direction), convection and radiation and q is specific heat source (heat density) at node of PV module.

Similarly, the energy balance equation for 'j' surface nodes of back panel is written as:

$$\Delta Q(zone)_{j,z} + \Delta Q(cond)_{j,j-1} + \Delta Q(conv)_{j,k} + \Delta Q(cond)_{j,j+1} + \sum_{i} Q(rad)_{j,i} = 0 \quad (2)$$

Where, $\Delta Q(zone)$ is the heat exchange term between the insulation panel and the room air zone. The energy balance equation for 'k' air nodes is written as:

$$\Delta Q(enth)_{k,k+1} + \Delta Q(conv)_{k,i} + \Delta Q(conv)_{k,j} = 0 \quad (3)$$

Where, ΔQ(enth) is the enthalpy change for the air nodes in the upward direction.

The solution of nodal equations of thermal network comprises formulation of conductance's matrix and heat source matrix [2]. Matrix equation has heat source vector terms for Q's and U's are the conductances at various nodes. The general form of thermal resistance network in matrix form is written as [2]:

$$U_{30,30} \times T_{30,1} = Q_{1,30} \quad (4)$$

The enthalpy change for an air node in the control volume with air node k and the downstream node being k-1 is equal to $U_{k,k-1}(T_k-T_{k-1})$. Taking into account the energy coming from the two solid surfaces (PV module and back panel), the following energy balance equation for the air nodes is obtained:

$$U_{k,k-1}\left(T_k - T_{k-1}\right) + U_{k,i}\left(T_k - T_i\right) + U_{k,j}\left(T_k - T_j\right) = 0 \quad (5)$$

Where, the conductance $U_{k,k-1}$ is equal to mc_p. The interpretation of the above equation in network modelling is that the conductance is positive in the upstream direction of energy flow and 0 in the downstream i.e. the air going through the air node in control volume is not affected by the downstream air node in previous control volume. The energy balance nodal equation for nodes of PV module, i nodes:

$$\left(T_i - T_o\right) U_i + \frac{q_i}{n} + U_{i,i+1}\left(T_i - T_{i+1}\right) + U_{i,i-1}\left(T_i - T_{i-1}\right) + U_{i,k}\left(T_i - T_k\right) + \sum_{j=20}^{29} U_{i,j}\left(T_i - T_j\right) = 0 \quad (6)$$

Where, U_i is the ambient heat transfer coefficient, $U_{i,i-1}$ and $U_{i,i+1}$ are conductances of PV module along the discretised height Δy. $U_{i,k}$ is the conductance value for convective heat transfer coefficient between the surface of PV module and air and $U_{i,j}$ is the radiative conductances between the PV module and back panel. Similarly energy balance nodal equation for nodes of insulation panel, j nodes:

$$\left(T_j - T_R\right) U_j + U_{j,j+1}\left(T_j - T_{j+1}\right) + U_{j,j-1}\left(T_j - T_{j-1}\right) + U_{j,k}\left(T_j - T_k\right) + \sum_{i=0}^{9} U_{i,j}\left(T_j - T_i\right) = 0 \quad (7)$$

Where, U_j is the heat loss coefficient for the insulation panel and room air zone, $U_{j,j-1}$ and $U_{j,j+1}$ are conductances of back panel along the discretised height Δy and $U_{j,k}$ is the conductance value for convective heat transfer coefficient between the surface of back panel and air.

The procedure for obtaining numerical solution of the conjugate heat exchange network is summarised as [2]:

Step 1. The input values as presented in Table II are used to initialise the simulation model.

Step 2. The constitutive relations for conduction, convection, radiation and heat transport are used to calculate conductance terms of the conjugate heat exchange network.

Step 3. The measured air velocity is used to obtain the mass flow rate for one PV module for calculating thermal capacity conductance for heat transport.

Step 4. The non-linearity of temperature dependent boundary conditions is eliminated by heat balance model.

Step 5. The effect of radiation heat exchange between surface nodes of PV module and insulation panel is considered in detail with radiosity-irradiation method assuming enclosure analysis.

Step 6. The convective heat transfer coefficients are calculated using temperatures obtained from the heat balance model.

Step 7. The calculated conductance terms are used to formulate the U-matrix (30 X 30) after obtaining off-diagonal and diagonal entries.

Step 8. The inverse of U-matrix (30 X 30) is multiplied with the heat source element matrix (1 X 30), to obtain temperatures at 30 nodes (30 X 1).

Results and Discussions

A full scale outdoor experimental test facility for a photovoltaic amplifier was set up at Concordia University, Montréal, Canada [2]. The experimental investigations were conducted to generate heat and D.C. electricity from a parallel plate photovoltaic (PV) amplifier. The measured thermal and electrical performances of a photovoltaic amplifier were established due to varying electrical resistance with rotation of knob of potentiometer without interrupting the current [2]. The tests were conducted for obtaining currents, voltages, heat generation viz., temperatures, air velocities, sensible heat, thermal time constants and thermal storage capacity of glass coated PV modules installed on a wooden frame [2]. The current-voltage measurements were obtained for determining electric power output with a series electrical circuit connection of a pair of vertically inclined PV modules installed on a wooden frame. The temperatures were measured as a function of volumetric capacity viz., height of a parallel plate PV amplifier. The non-linear thermal results include measurements of temperatures for PV modules, insulating panel and ventilated air column in the wooden frame. The air velocities were developed in the ventilated air column for transportation of heat both as a measure of buoyancy and fan induced hybrid ventilation. The measured electrical results of currents, voltages and power with varying electrical resistance of potentiometer are presented in Table I [6]. The measured thermal results of temperatures of various components of PV device, ambient air and room air temperatures, air velocities and solar intensities are presented in Table II. The results of the power output from a potentiometer with rotation of its circular knob are illustrated in Figure 3 [6]. The sample results that were obtained from the experimental setup of a photovoltaic amplifier are presented in Table II. The location of sensors as per Figure 1 is presented in Table III. The measurements that were obtained from a photovoltaic amplifier are for successive runs of solar ventilation at fluctuating hybrid air velocities. The measurements were collected for a pair of successive runs at same solar intensities. The experimental results from a photovoltaic amplifier are plotted in Figure 4(a) and Figure 4(b). The air temperatures in the solar wall are developed as a function of magnitude of the ambient air temperature. The maximum convection heat transfer was observed between the ambient cold air entering into the inlet damper and the air passage of the test section. The dissipation of heat at the inlet opening has resulted in the considerable increase in temperature of the air. As shown in Figure 4(a) with ambient air temperature around 15°C, mean temperature of PV module was varying between 35°C and 38°C and mean temperature of insulation panel was 25°C and air temperature near the outlet of PV module test section was between 20-22°C [2]. The effect of absorbed solar radiation on temperature difference with height for air, back panel and PV module is shown in Figure 4(b).The temperature difference between top and bottom node for PV module was around 2°C, for insulation panel between 8-10°C and temperature difference between ambient air and air leaving the air velocity sensor in a parallel plate air passage was between 5-9°C [2]. The insulation panel was heated by long wave radiation heat exchange from PV module and its temperature at top node was 15°C higher than ambient air and surface temperature difference between top and bottom nodes of the insulation panel was 10°C [2]. This was because minimum long wave radiation heat exchange between parallel plates of a photovoltaic amplifier at their top nodes and maximum heat transfer to air flowing through the air passage was at bottom nodes [2]. This was the cause for increased temperature difference between top and bottom nodes of the insulation panel [2].

The average air velocity of six scanned readings taken every two minutes between 11:00 a.m. and 12:30 p.m. was varied between 0.4 m/s and 0.65 m/s, whereas the average value of the whole data of readings was 0.52 m/s [2]. It was validated by one of the experiments that measured surface temperatures of the two adjacent PV modules were within ± 0.5 °C under similar operating conditions [2]. The results obtained from computer simulation model for Run no. I of Table II are plotted in Figure 4(a) to Figure 4(c). The temperature profiles of PV module, air and insulation panel are obtained with respect to the height of PV module test section. The initial boundary values in the model are based on the measured values of quasi steady-state solar irradiation, ambient air temperature, and room air temperature and air velocities. The measured boundary values were then used to obtain solutions for heat equations representing heat exchange and transport in a photovoltaic amplifier. The results of the simulation model are obtained by assuming constant air and surface properties. The mean air velocity was assumed be constant in the test section for representing nodes of air with same thermal capacity conductances, mc_p also termed as thermal capacity rate C_p through out its height [2]. The amount of dissipated heat from the PV module taken up by air is dependent on mass flow rate passing over it, thermal capacity of air and temperature difference between air-column in the test section [2]. The thermal capacity rate mc_p is ventilation heat capacity developed for width W of single PV module due to dissipated heat [2]. The dissipated heat from the face area (W X H) of PV module is ventilation heat rate is convection heat loss from the walls of PV module test section due to difference in temperature between the walls of test section and air flowing across cross-sectional area of test section across one PV module [2]. The simulated temperatures are fitted with their regression values from their ideal exponential or logarithmic behaviour as shown in Figure 4.

Table I Electrical measurement results with varying resistance of potentiometer

Rotation	Volts	Amps	Watts	Rotation	Volts	Amps	Watts	Rotation	Volts	Amps	Watts
240°	18.7	-	-	83°	16.3	0.935	15.23	30°	9.7	1.577	15.21
239°	16.5	0.331	5.461	75°	16.0	1.014	16.26	27°	9.0	1.587	14.33
201°	17.4	0.414	7.195	69°	15.8	1.100	17.38	21°	7.1	1.583	11.24
185°	17.5	0.454	7.940	64°	15.5	1.165	18.04	18°	6.2	1.573	9.831
162°	17.3	0.513	8.885	55°	15.0	1.302	19.53	17°	5.7	1.578	9.026
150°	17.18	0.550	9.449	50°	14.5	1.386	20.05	12°	3.9	1.567	6.257
142°	17.19	0.582	10.00	43°	13.2	1.503	19.79	10°	3.2	1.553	4.840
128°	17.1	0.640	10.93	42°	13.1	1.493	19.49	1.5°	0.5	1.593	0.807
107°	16.8	0.750	12.51	37°	11.9	1.536	18.26	1°	0.3	1.59	0.426
89°	16.4	0.884	14.45	32°	10.5	1.567	16.42	0°	-	1.643	-

Table II Sample experimental results obtained from a photovoltaic amplifier

Run No.	S ($W m^{-2}$)	E_p (W)	T_o (° C)	T_s (° C)	V ($m s^{-1}$)	$T_p(b)$ (° C)	$T_p(m)$ (° C)	$T_p(t)$ (° C)	$T_b(b)$ (° C)	$T_b(m)$ (° C)	$T_b(t)$ (° C)	$T_a(m)$ (° C)	$T_a(t)$ (° C)
I	725.4	31.1	14.5	22.6	0.425	34.2	32.9	36.3	19.9	24.0	29.4	19.2	22.4
II	725.4	31.1	15.9	22.9	0.362	32.5	33.3	35.7	20.2	23.9	29.1	19.7	23.3

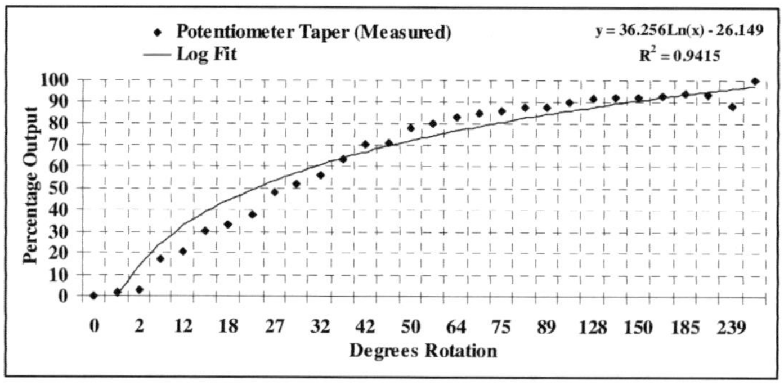

Figure 3. Potentiometer Taper (Measured) with percentage voltage output

Table III Location of sensors in the experimental setup

Distance as per location of sensors (Figure 1)	$T_p(b)$ (° C)	$T_p(m)$ (° C)	$T_p(t)$ (° C)	$T_b(b)$ (° C)	$T_b(m)$ (° C)	$T_b(t)$ (° C)	$T_a(b)$ (° C)	$T_a(m)$ (° C)	$T_a(t)$ (° C)	Air velocity sensor
y (cm)	15	55	94	15	55	94	15	55	94	99
z (cm)	60	60	60	60	60	60	60	60	60	60
x (mm)	6.2	6.2	6.2	96.2	96.2	96.2	51.2	51.2	51.2	51.2

Note: x is horizontal; y is vertical; z is adjacent 3rd axis of x-y plane of Figure 1

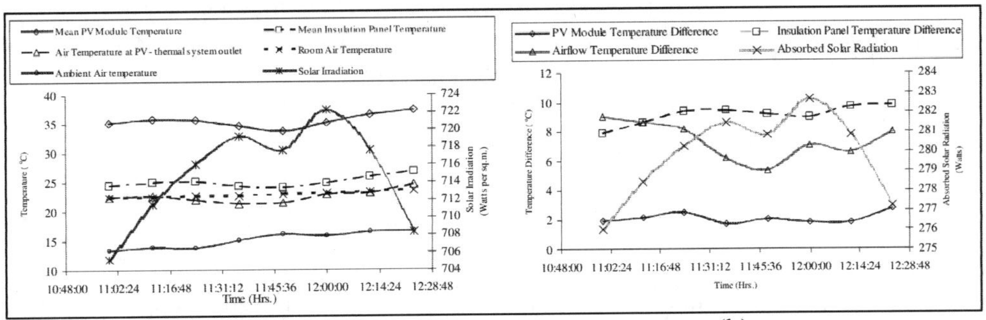

(a) (b)

Figure 4. Experimental results of a photovoltaic amplifier: (a) Mean temperatures of PV Module, air and back panel; and (b) Temperature differences along the height

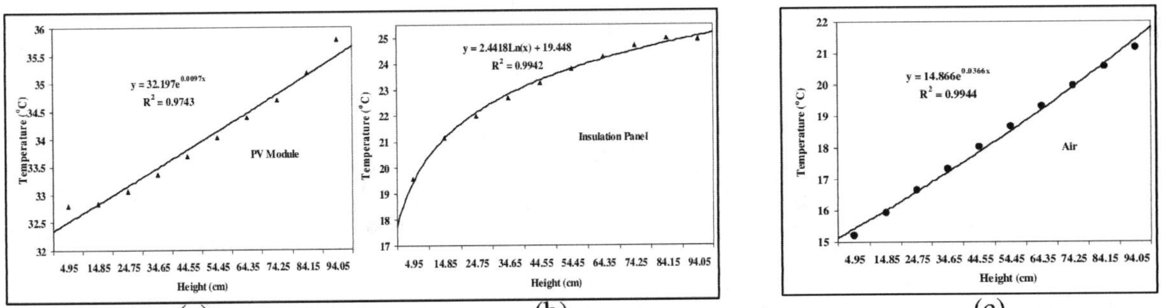

(a) (b) (c)

Figure 5. Computer simulation results: (a) PV module temperature; (b) insulation panel temperature; and (c) air temperature

Conclusive Remarks

The coordination of the activities of the various sensory organs within the body is controlled by the secretion of hormones and by transmission of nerve impulses. The nerve impulses travel along axons. The sensory nerve axons carry impulses from different parts of the body to the central nervous system (brain and spinal cord). Although parts of the brain involved in various physiological functions have been located, research attempts to localize complex behaviour patterns in sensory areas of brain have failed [1]. The sensory areas are located for the impulses on the cortex. The motor areas are also located where "command" impulses leave the cortex. The cerebellum receives impulses from sensory receptors and interacts with motor cortex, insuring smooth muscular movements. The reticular formation receives sensory impulses from the environment and sends impulses to the cortex.

An example of a parallel plate photovoltaic amplifier is presented to illustrate its complex behaviour as a sensor and transducer for noise filtering. The phenomenon of photovoltaic amplification was detected from the graphs of Figures 3 and 5. The gain in steady state electrical and thermal functions for a photovoltaic device was a factor of its volume or resistance. The non-linear thermal results of temperatures for PV modules, insulating panel and ventilated air column were measured as a function of volume of a parallel plate photovoltaic amplifier. The air velocities were developed in the ventilated air column of a photovoltaic amplifier for transportation of heat.

A simulation model for prediction of temperature distributions varying with volume of a parallel plate photovoltaic device was developed. The model was used to predict temperature distributions at pre-defined locations in PV module, plywood board and air flowing through parallel plate geometry. The model is a useful tool for waste heat recovery design for obtaining necessary cooling of a photovoltaic amplifier whilst achieving heating, ventilating and air-conditioning (HVAC) requirements for the internal environment. The analysis of a parallel plate photovoltaic amplifier connected to a potentiometer has presented the cases of series and parallel cases of L-C-R circuit resonance [3]. The cases of resonance were easily analysed with the aid of presented experimental and modelling results. The series resonance represents the propagation of elastic waves of mechanical system of a photovoltaic amplifier. The parallel resonance occurs through propagation of heat waves and fluid surface waves in conjunction with inductance due to mass of a photovoltaic amplifier and resistance due to temperatures of ground surface and ambient air.

Acknowledgements

Dr. Himanshu Dehra is thankful to his wives Deepika (Queen of Spades), Jyoti (Queen of Clubs), Shweta (Queen of Hearts), Monica (Queen of Diamonds), and Kiran (Princess) for their love, loyalty, devotion and their psycho-social affiliation with **Mate Selection Methods** dated February 28, 2009.

References

1. Macey, Robert I. (1968), "Human Physiology", Printice Hall: Englewood Cliffs, New Jersey, USA.
2. Dehra H. (2004), "A Numerical and Experimental Study for Generation of Electric and Thermal Power with Photovoltaic Modules Embedded in Building Façade", submitted/unpublished Ph.D. thesis, Department of Building, Civil and Environmental Engineering, Concordia University, Montréal, Canada, August 2004.
3. Dehra H. (2007), "A Unified Theory for Stresses and Oscillations", Canadian Acoustics, **35**(3): 132-133.
4. Del Toro V. (1986), "Electrical Engineering Fundamentals", Prentice-Hall of India: New Delhi, India.
5. Dehra H. (2008), "The Noise Scales and their Units", Canadian Acoustics, **36**(3): 78-79.
6. Dehra H. (2008), "Power Transfer and Inductance in a Star Connected RC Circuit Amplifier", in the proceedings of American Institute of Chemical Engineers - AIChE 2008 Spring Meeting, New Orleans, LA, USA, April 6-10, 2008, session 96a.

Process Variability: Causes, Costs and Cures

Paper 6c
AIChE 2009
Spring National
Meeting

Doug White
Emerson Process
Management

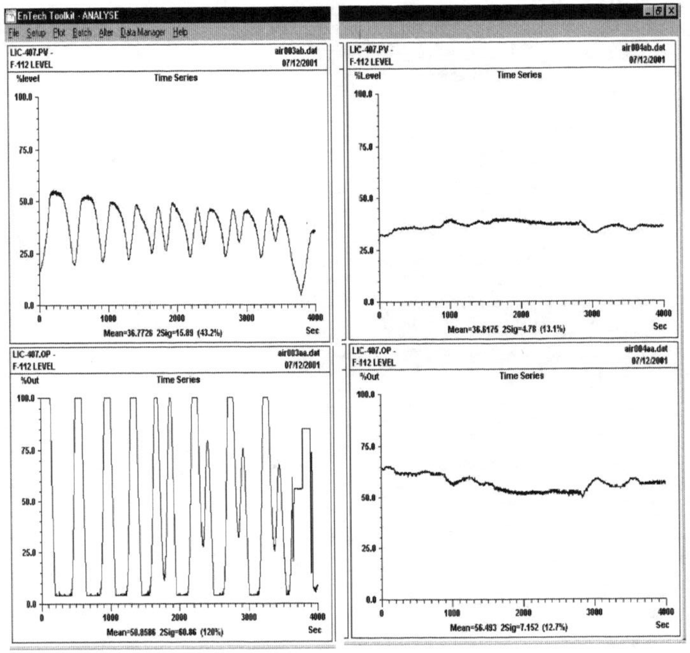

Speaker

Doug White

Principal Consultant and Vice President, APC Services

Advanced Applied Technologies

Process Systems and Solutions

Emerson Process Management

EMERSON Process Management

Many years experience designing, justifying, installing and commissioning advanced real time software applications in the process industries.

Session Objective

Help you recognize:

1. **The effects of excessive process operational variability in your plant – and the value of its reduction.**
2. **How you can reduce this variability through enhanced automation, advanced automation and asset management**

Agenda

- → Where does process variability come from?
- → Why do we care about process variability?
- → How can we solve variability problems?
- → How does excessive process variability affect the plant financial performance?
- → Case studies
- → What is the proper methodology to identify variability and justify variability reduction projects within the plant?

What are the types of Process Operational Variability?

Process Operational Variability

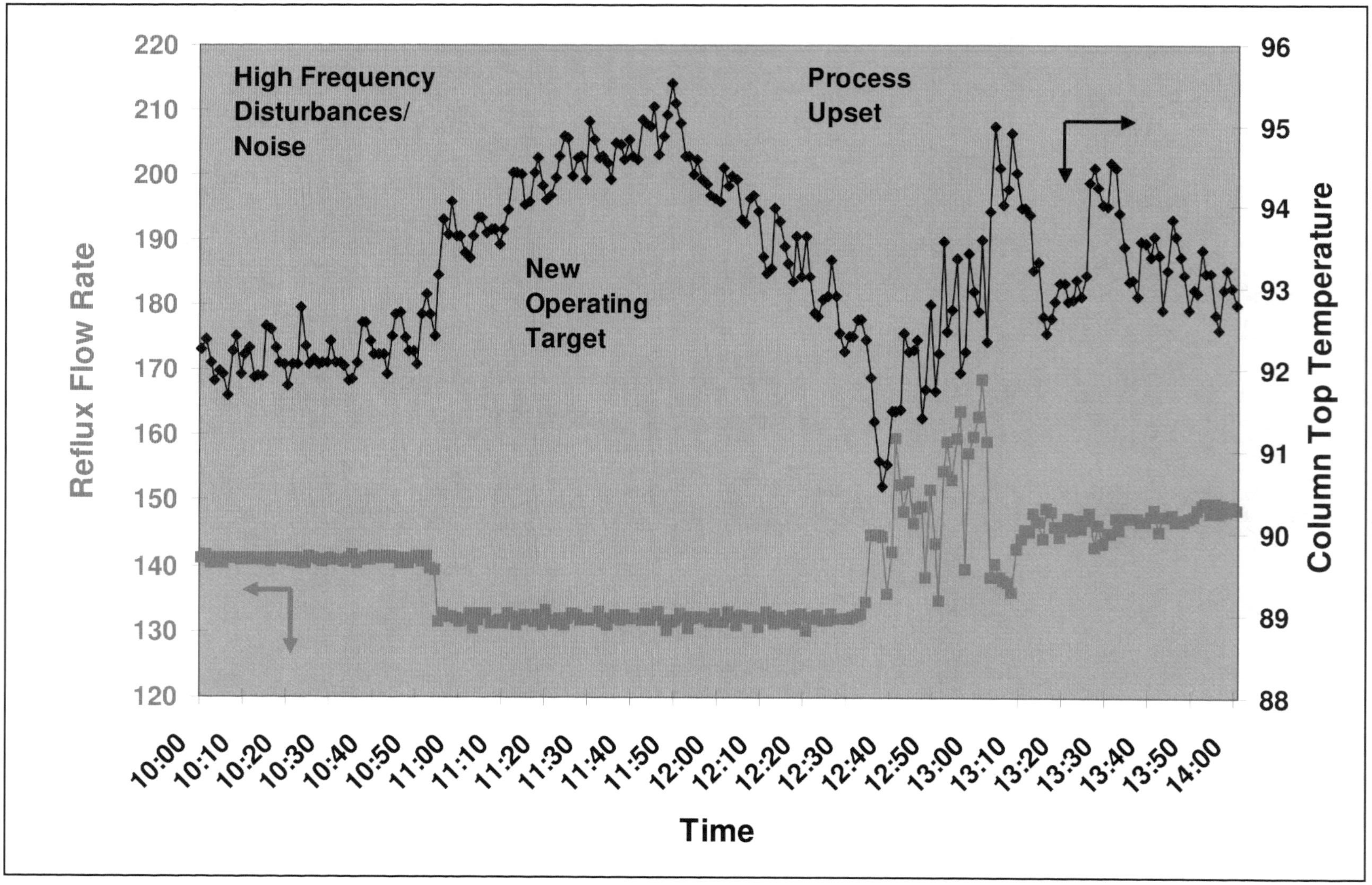

Types of Variability

Random Variability – High frequency variability, can be natural variation in materials, equipment or the process - seen as *noise* inherent to the process / measurement or *high frequency disturbances*

Cyclic Variability - often related to *process characteristics* or *control loop* performance

Process Upsets - sudden *disturbances (low frequency)* somewhere in the process

- "Load" changes – i.e. fuel gas quality, outside temperature, feed composition
- Setpoint Changes - new operating *targets,* i.e. feed rate, changed grade or mode of operation
- Batch/ Discontinuous Operation - New batch or grade change

Example – Noise/ High Frequency Disturbance

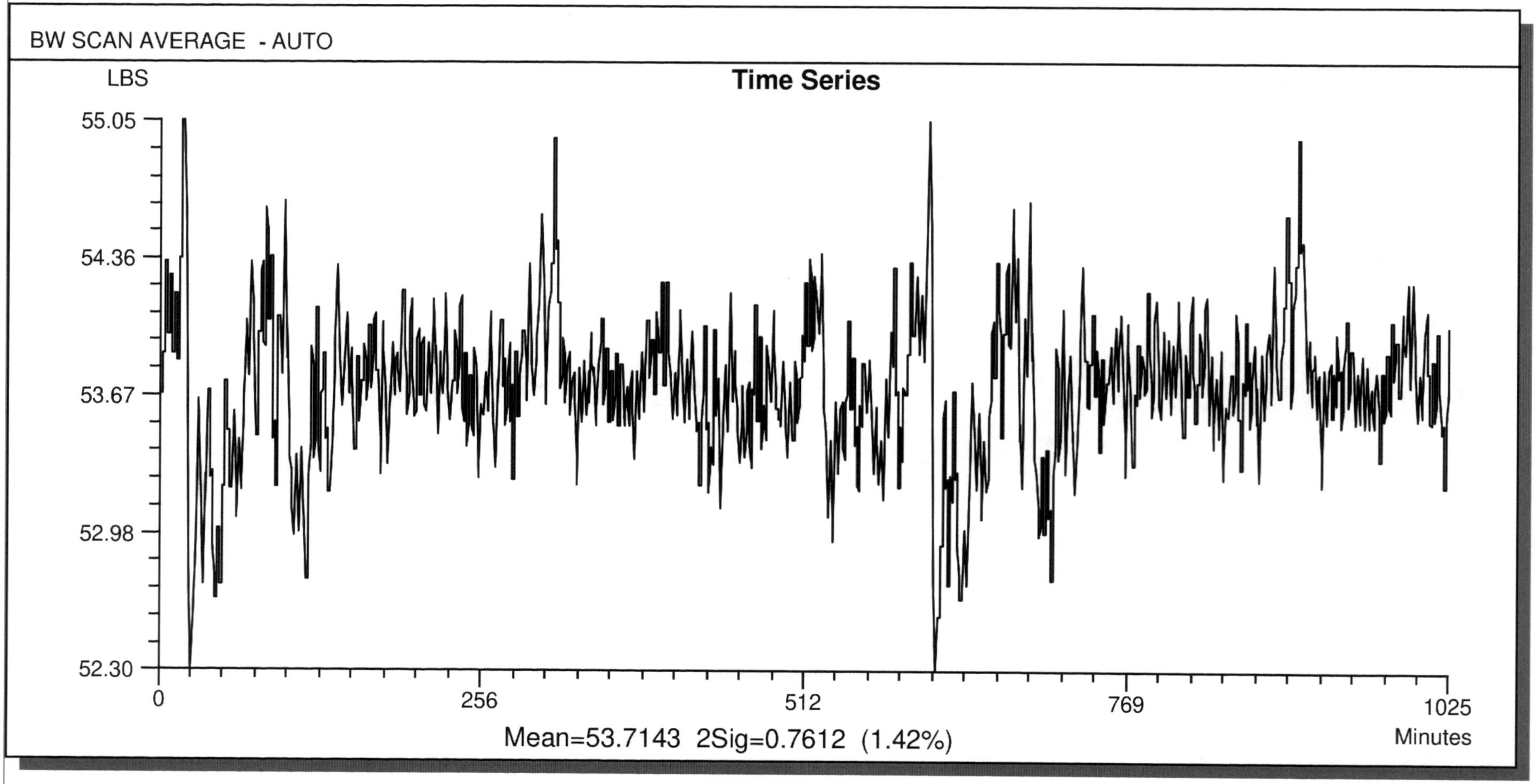

Example – Cyclic Variability

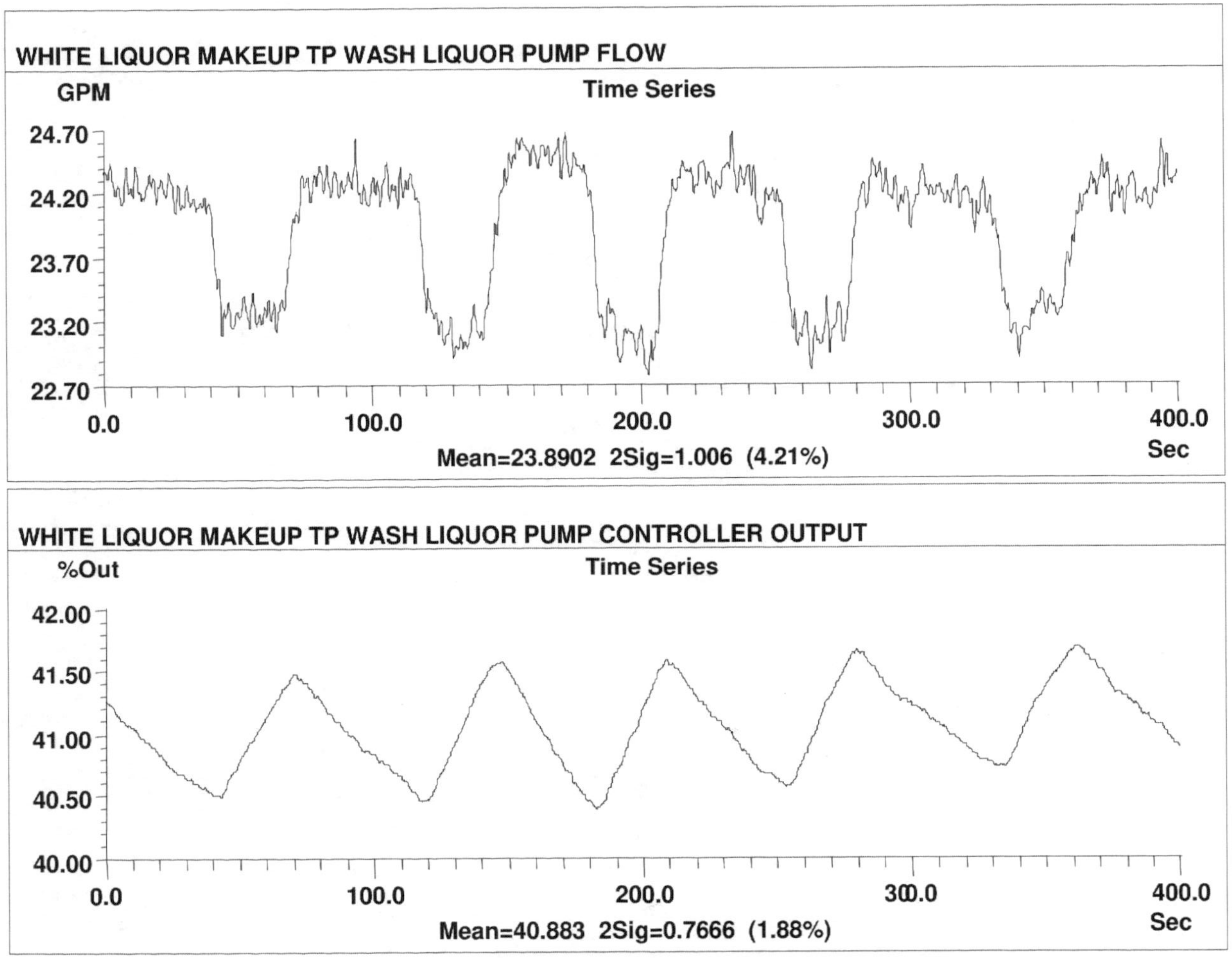

Very Common – Often Induced by Valve /Process /Controller Interaction

Example – Load Change

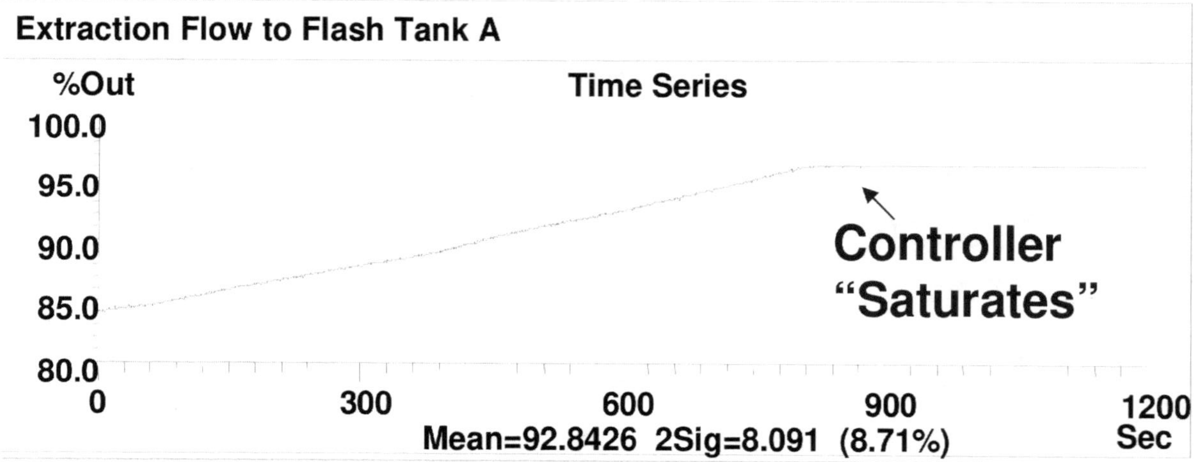

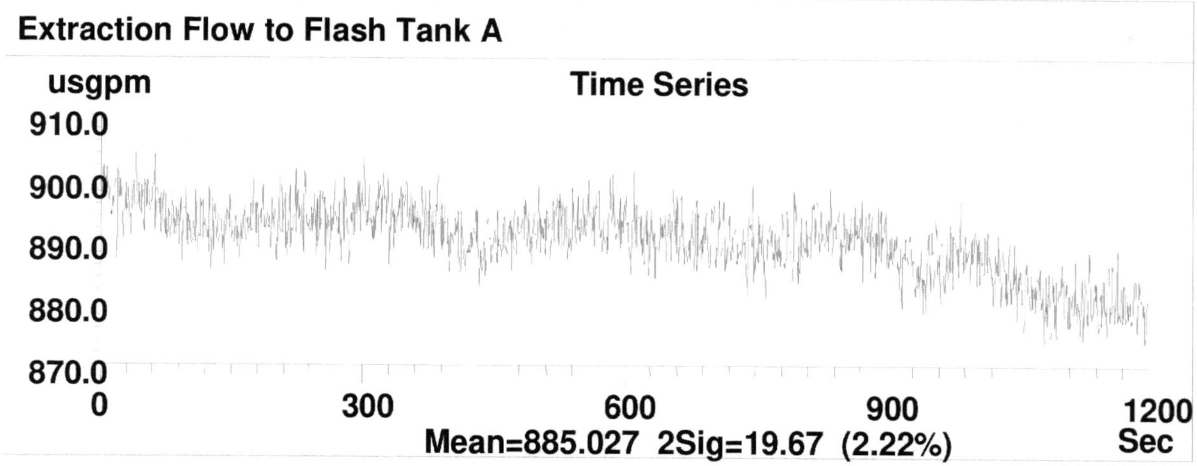

Batch Reactor Variability

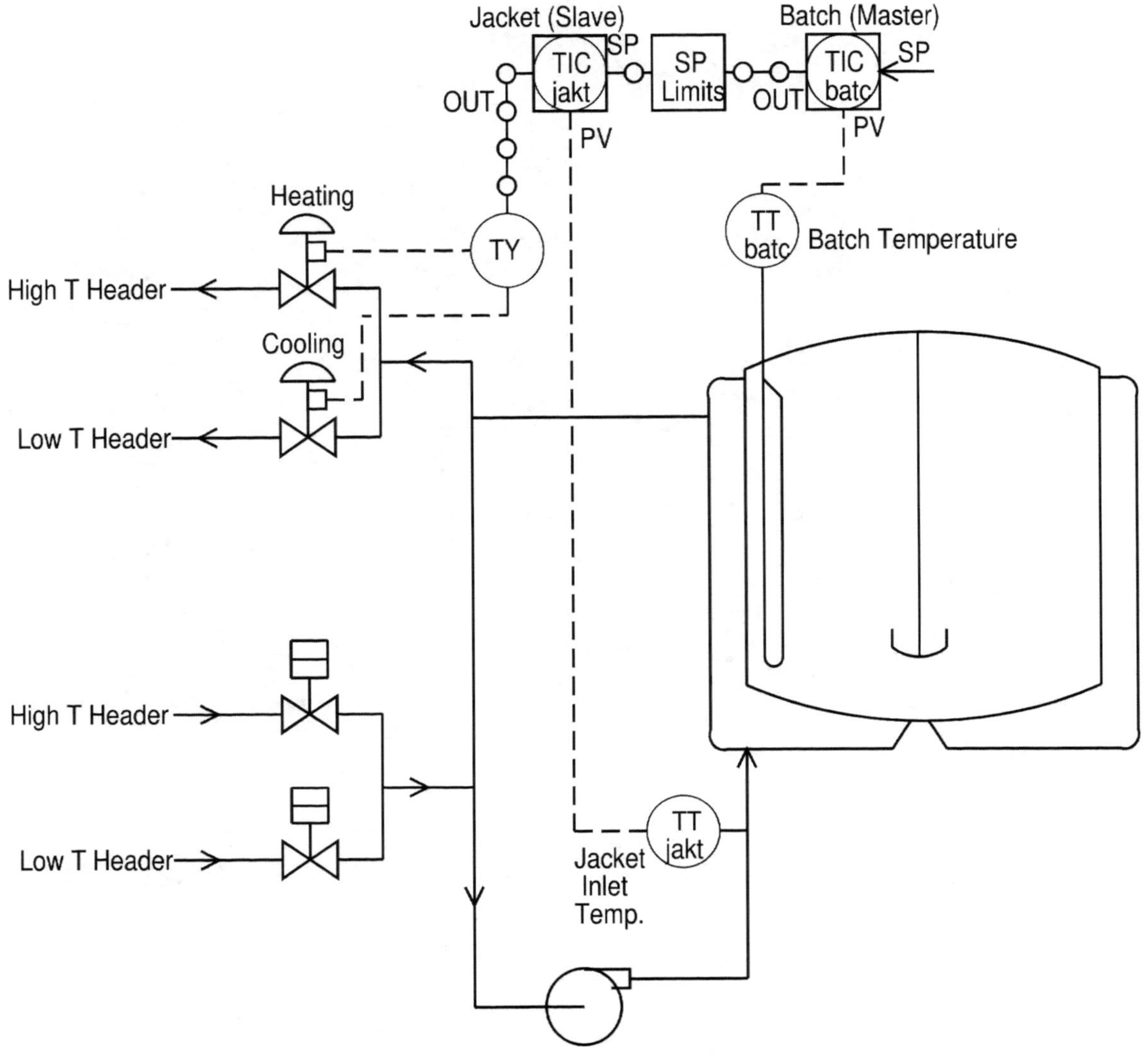

Batch Plant Variability – Total Batch Time

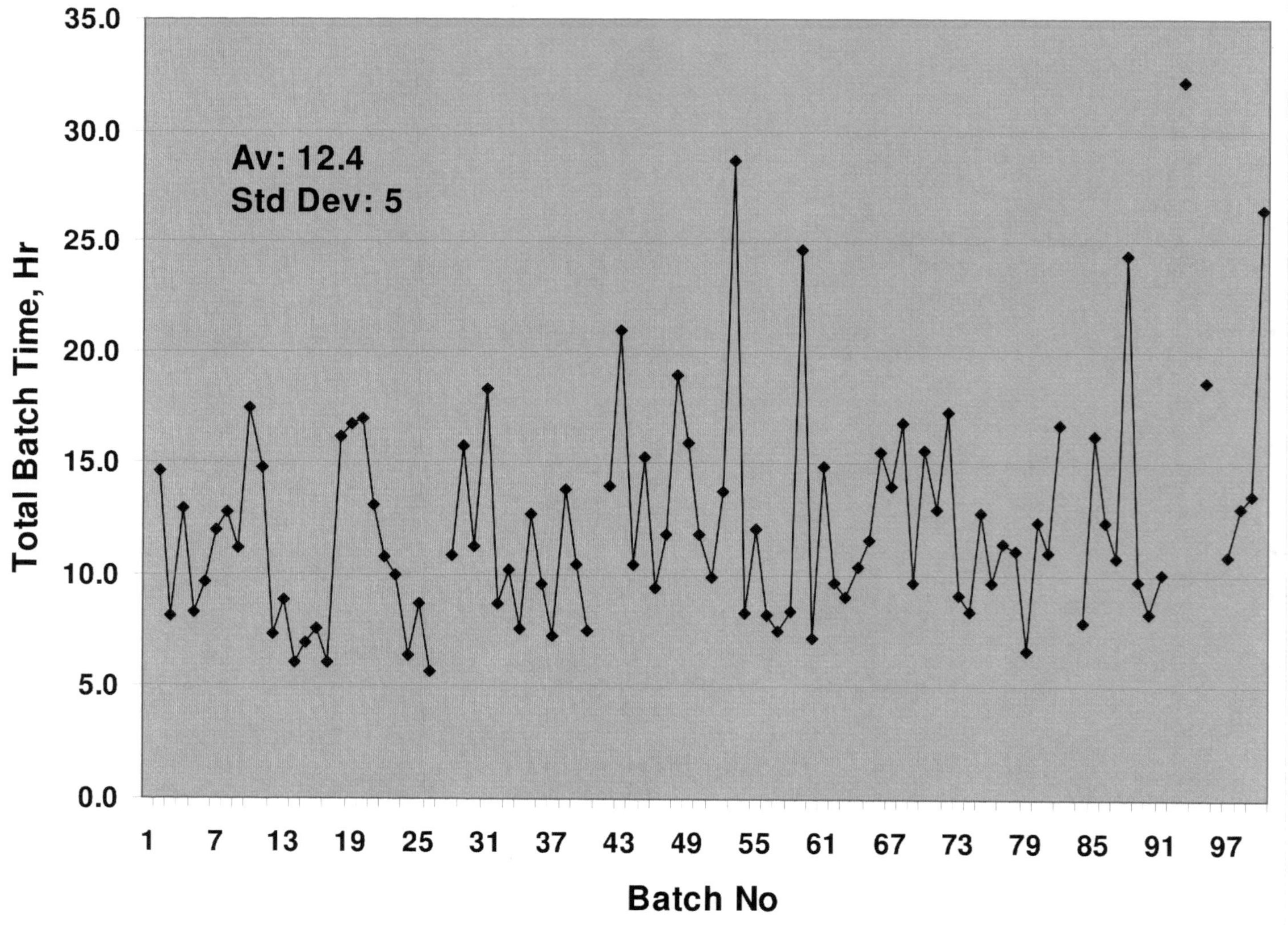

How Does The Process Propagate Operational Variability?

Variability Propagation – Process Integration

- **In some cases inlet and outlet streams are connected via heat exchangers to maximize thermal efficiency**
  - Provides heat interaction between inlet and outlet streams
  - Tends to propagate variability by causing incoming disturbances to impact entire unit operation

NGL Fractionation Demethanizer - Heat Interaction

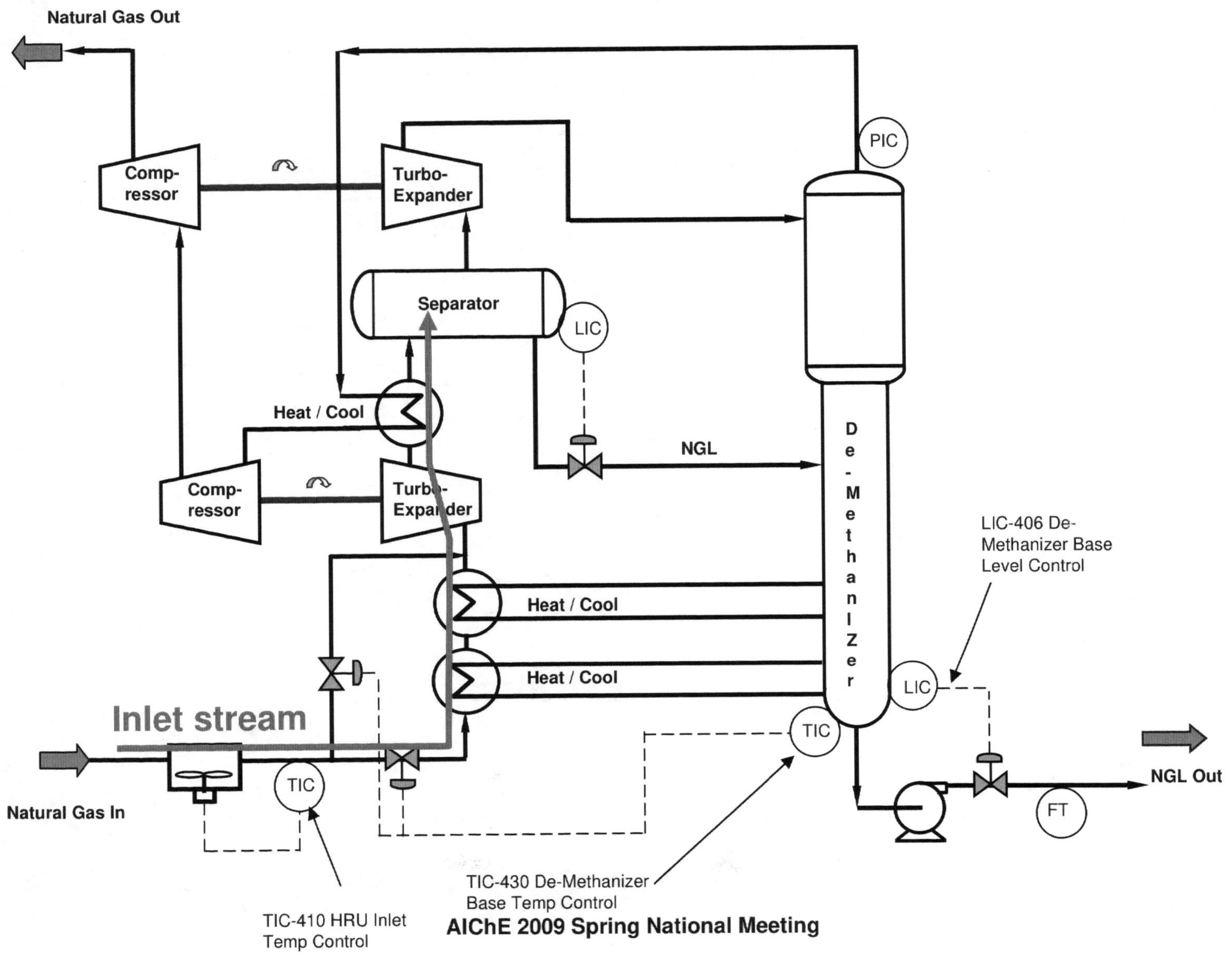

AIChE 2009 Spring National Meeting

NGL Fractionation Demethanizer - Heat Interaction

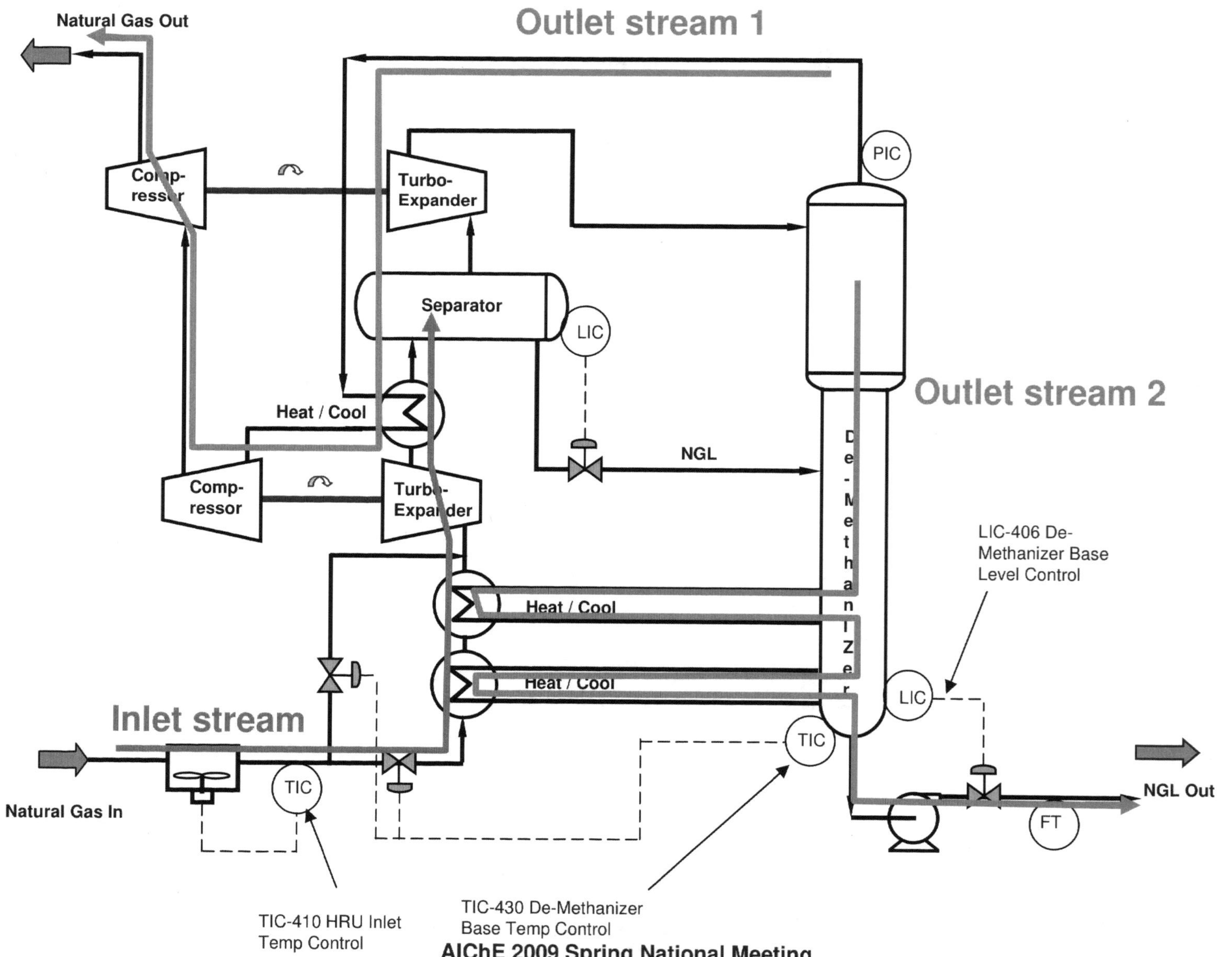

AIChE 2009 Spring National Meeting

NGL Fractionation Demethanizer - Heat Interaction

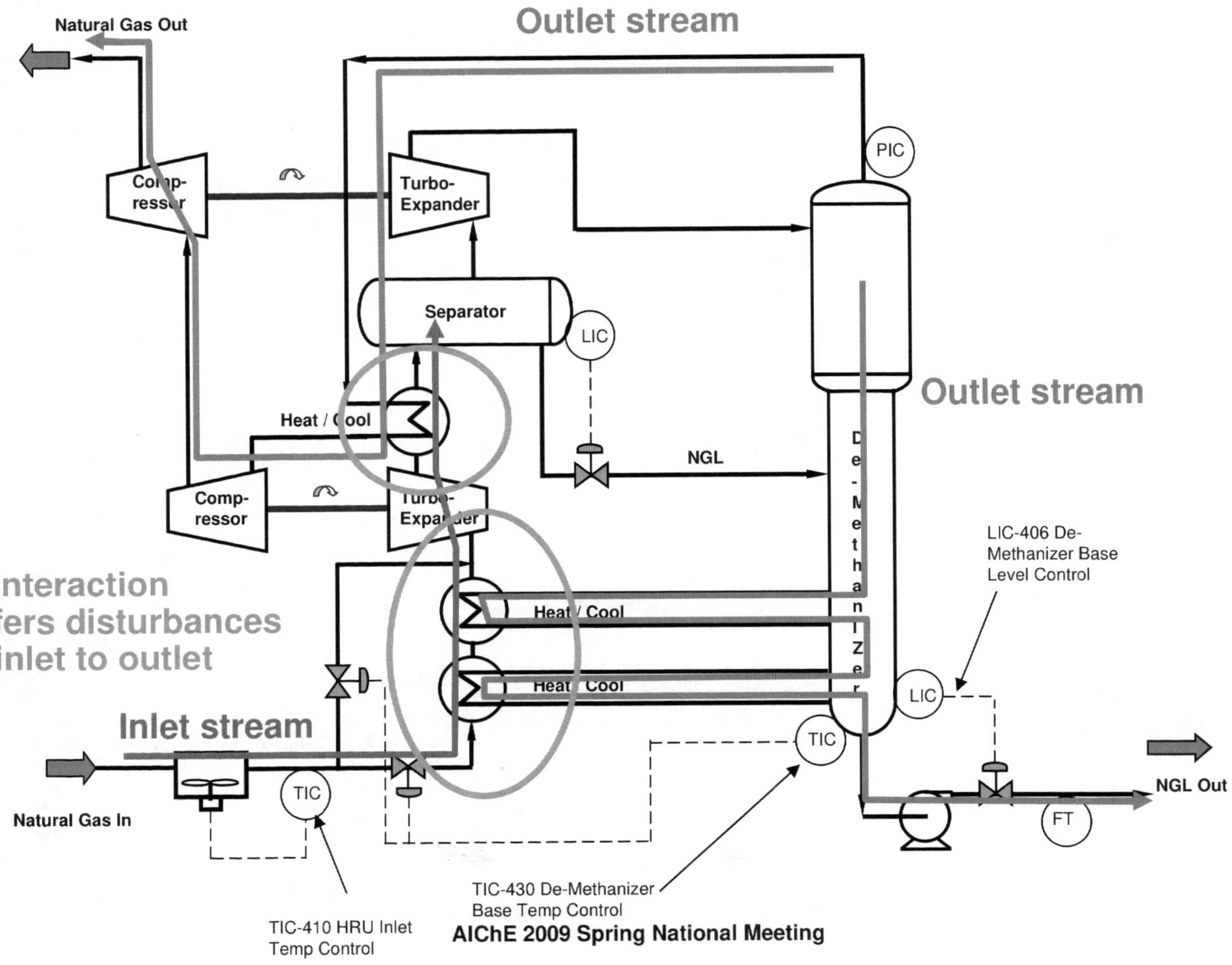

NGL Fractionation Demethanizer - Effect of Process Integration

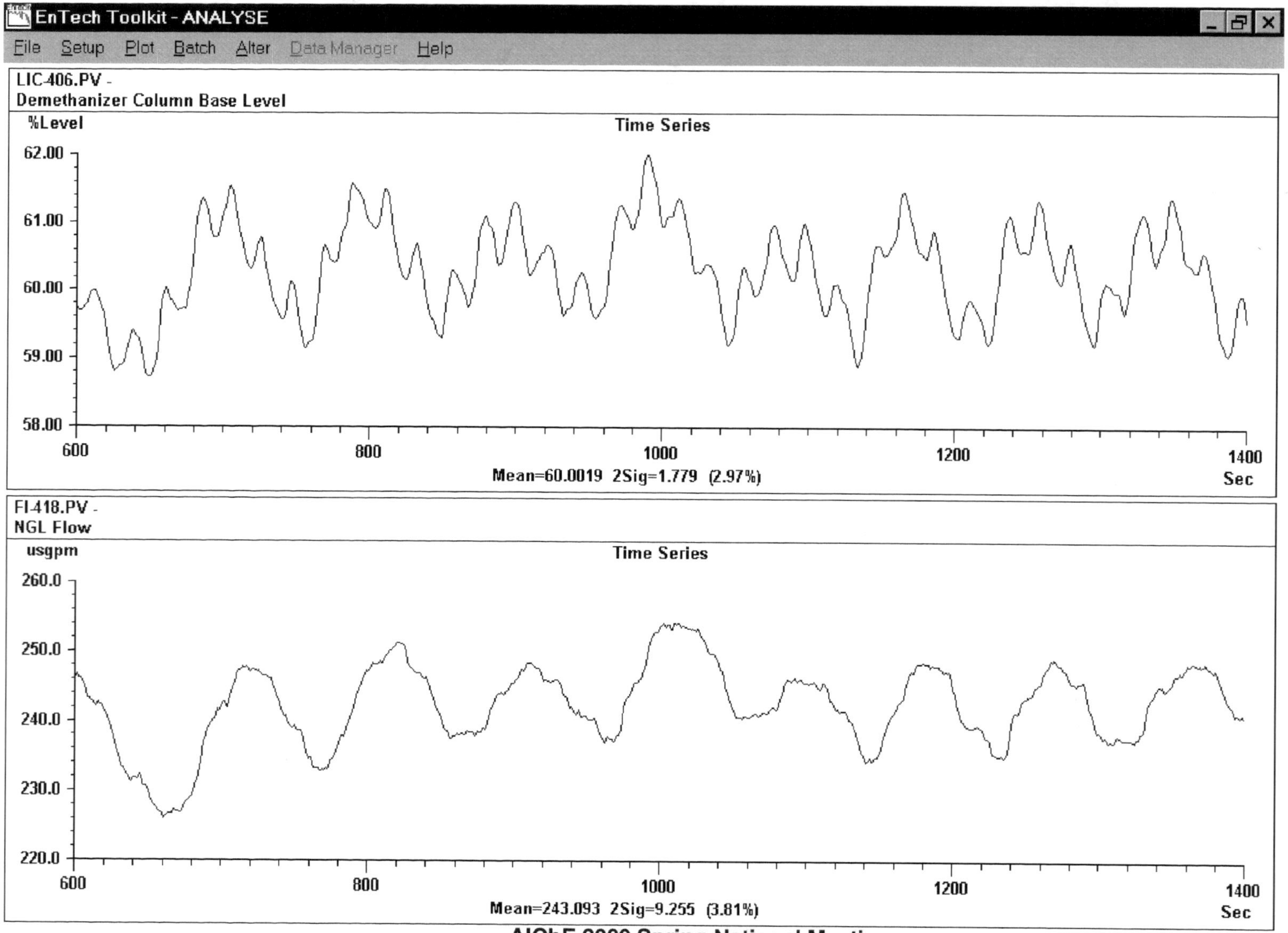

Variability Propagation – Process Recycle

- In some cases recycle of material or energy occurs in the process
  - Provides increased inventory hold-up and appears like a larger capacity
  - May be beneficial for smoothing out faster composition and temperature variability
  - Tends to complicate process dynamics making control much more difficult
  - Causes disturbances to persist much longer

Refining - Main Crude Unit - Recycle

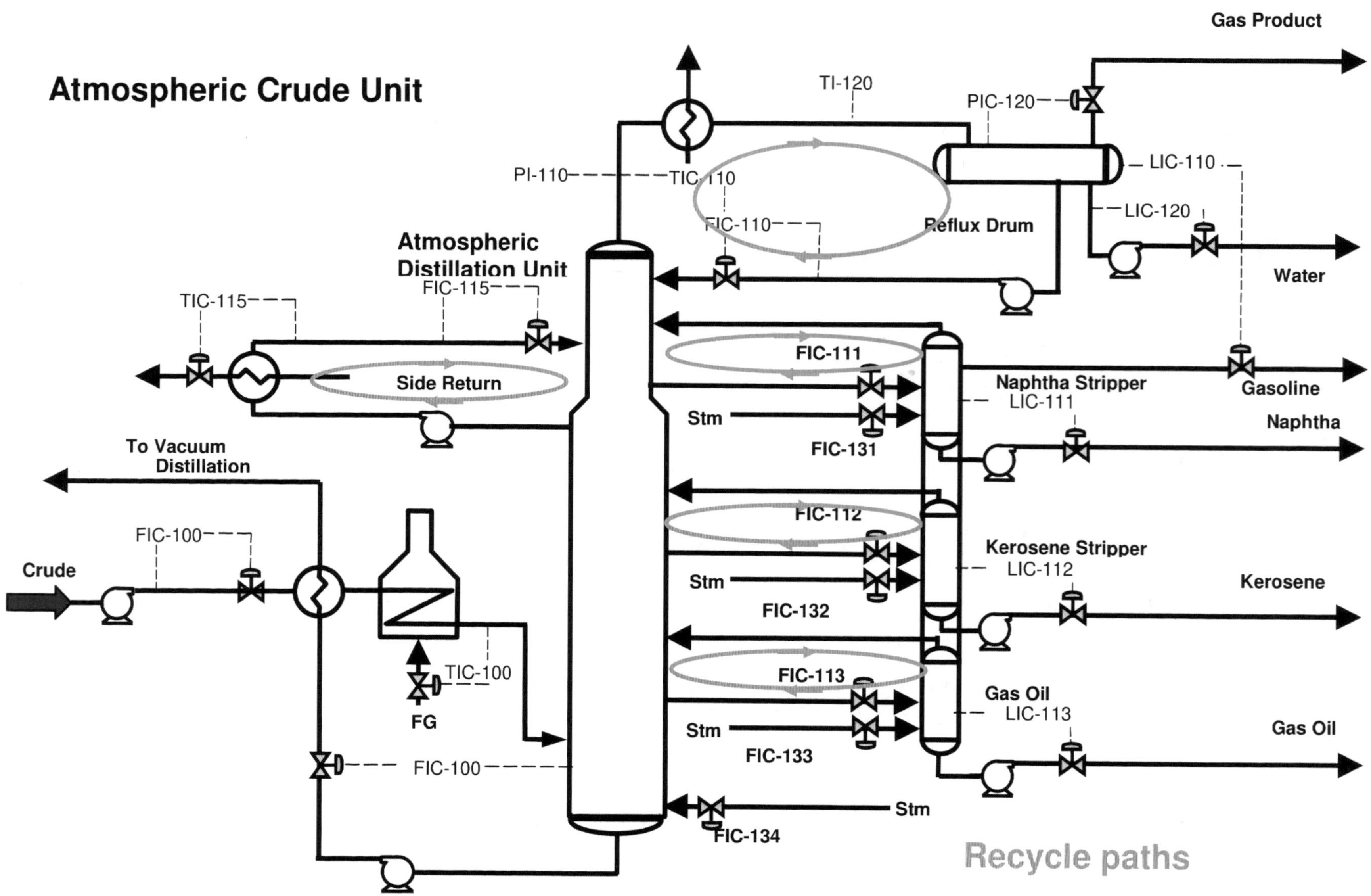

AIChE 2009 Spring National Meeting

Crude Unit Product Quality Variability

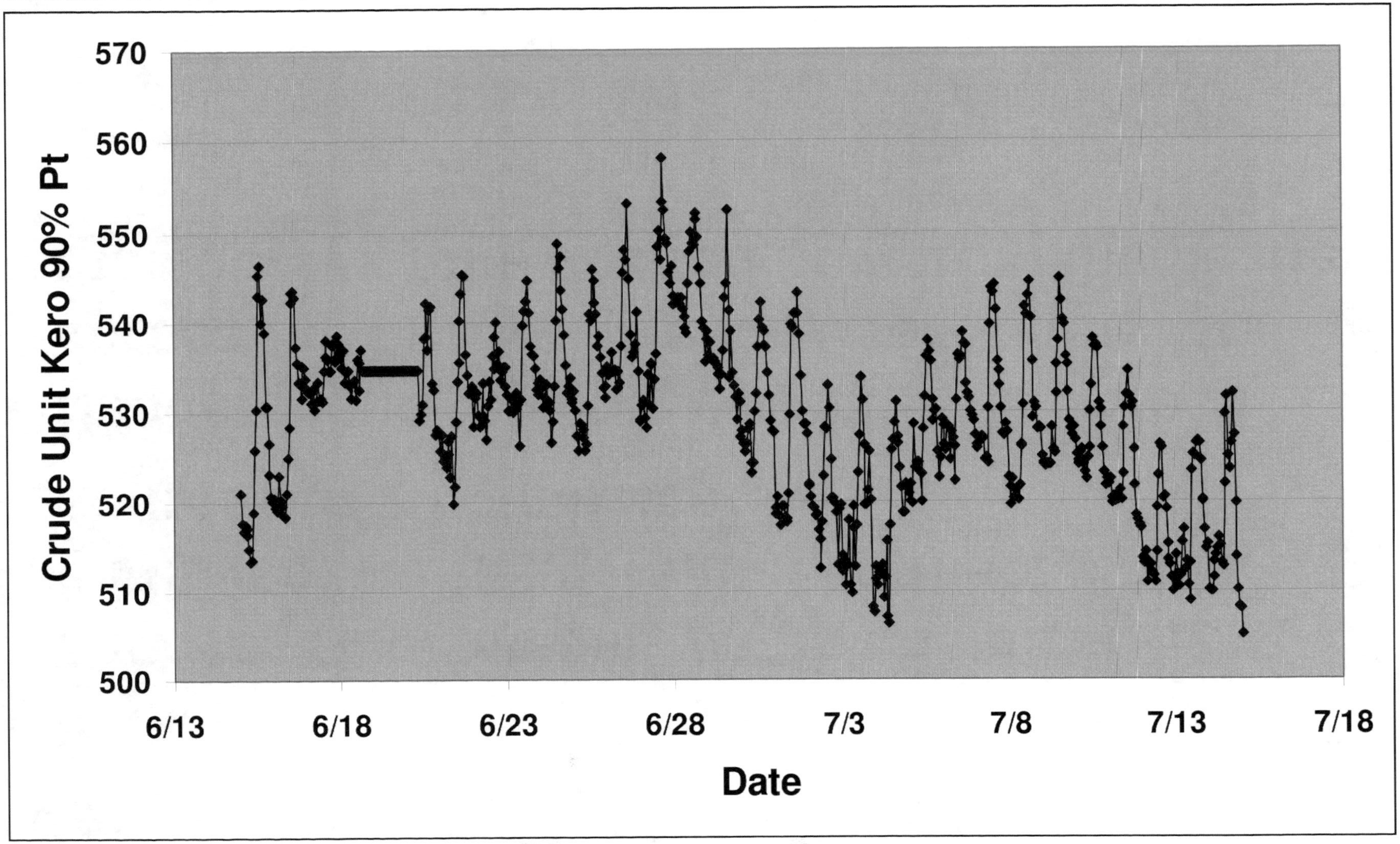

AIChE 2009 Spring National Meeting

Variability Propagation - Headers

- In many cases processes supply various services (fuel, steam, water, chemical additives) via headers that connect single sources to multiple users
- Headers tend to allow variability to spread across the header as each user “fights” for the supply of the desired flow from the header

Hydrocracker Quench Interaction

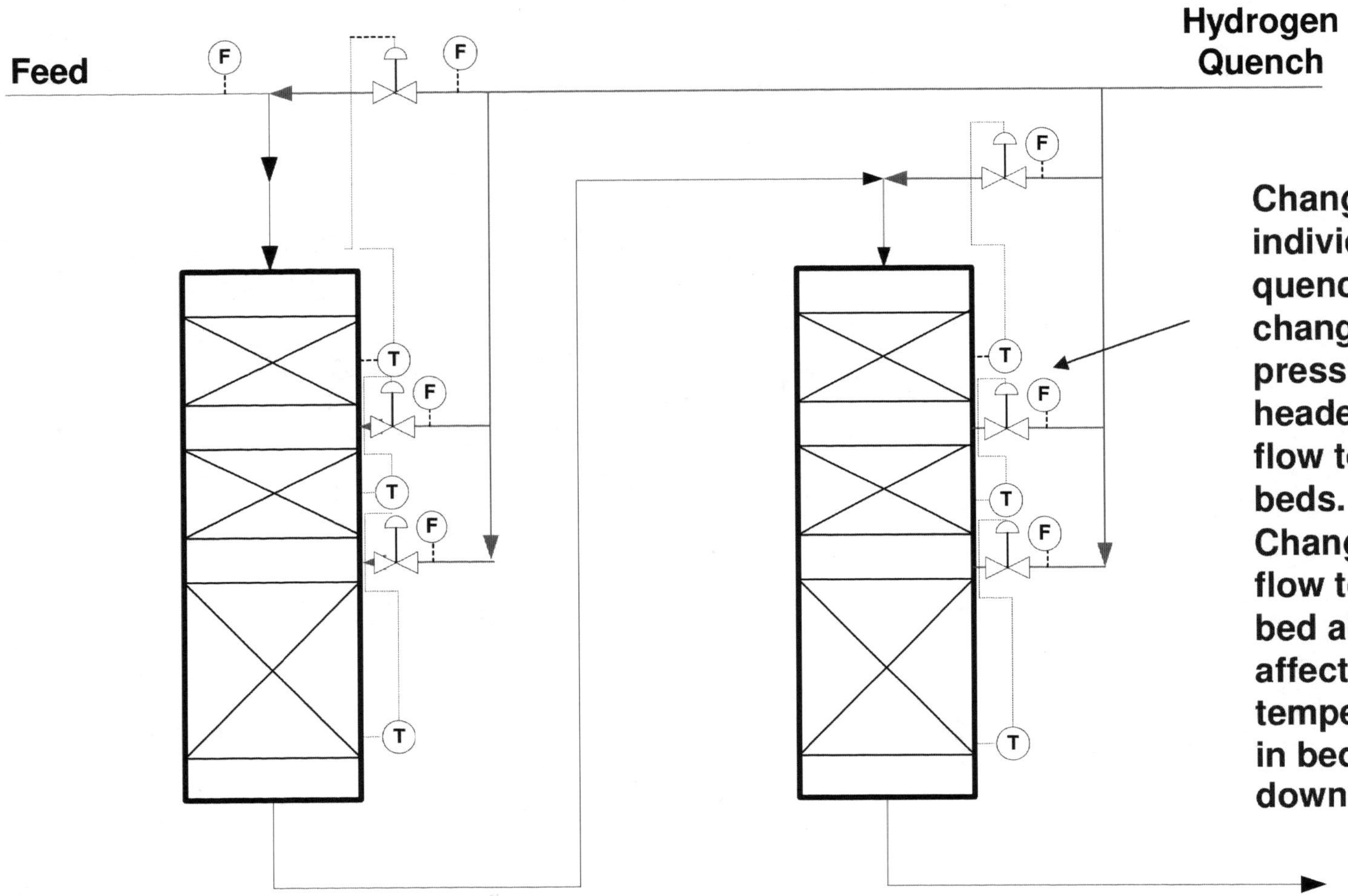

Change in individual quench flow changes pressure in header – and flow to other beds.
Change in flow to one bed also affects temperature in beds downstream.

Hydrocracker – Temperature Control

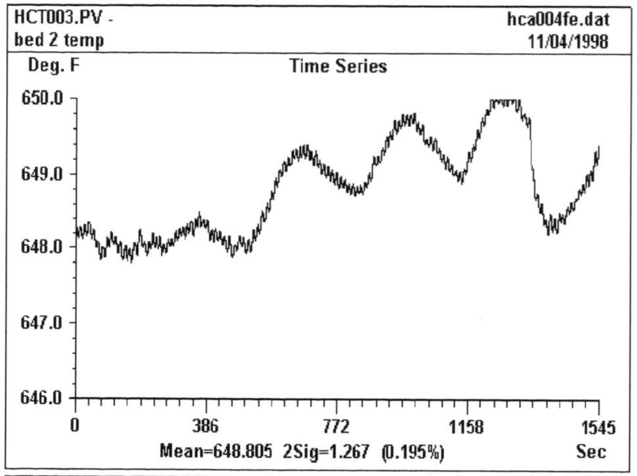

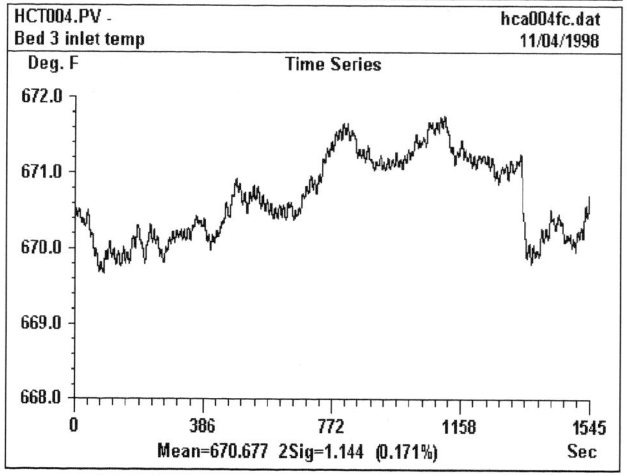

AIChE 2009 Spring National Meeting

Steam Header Interaction

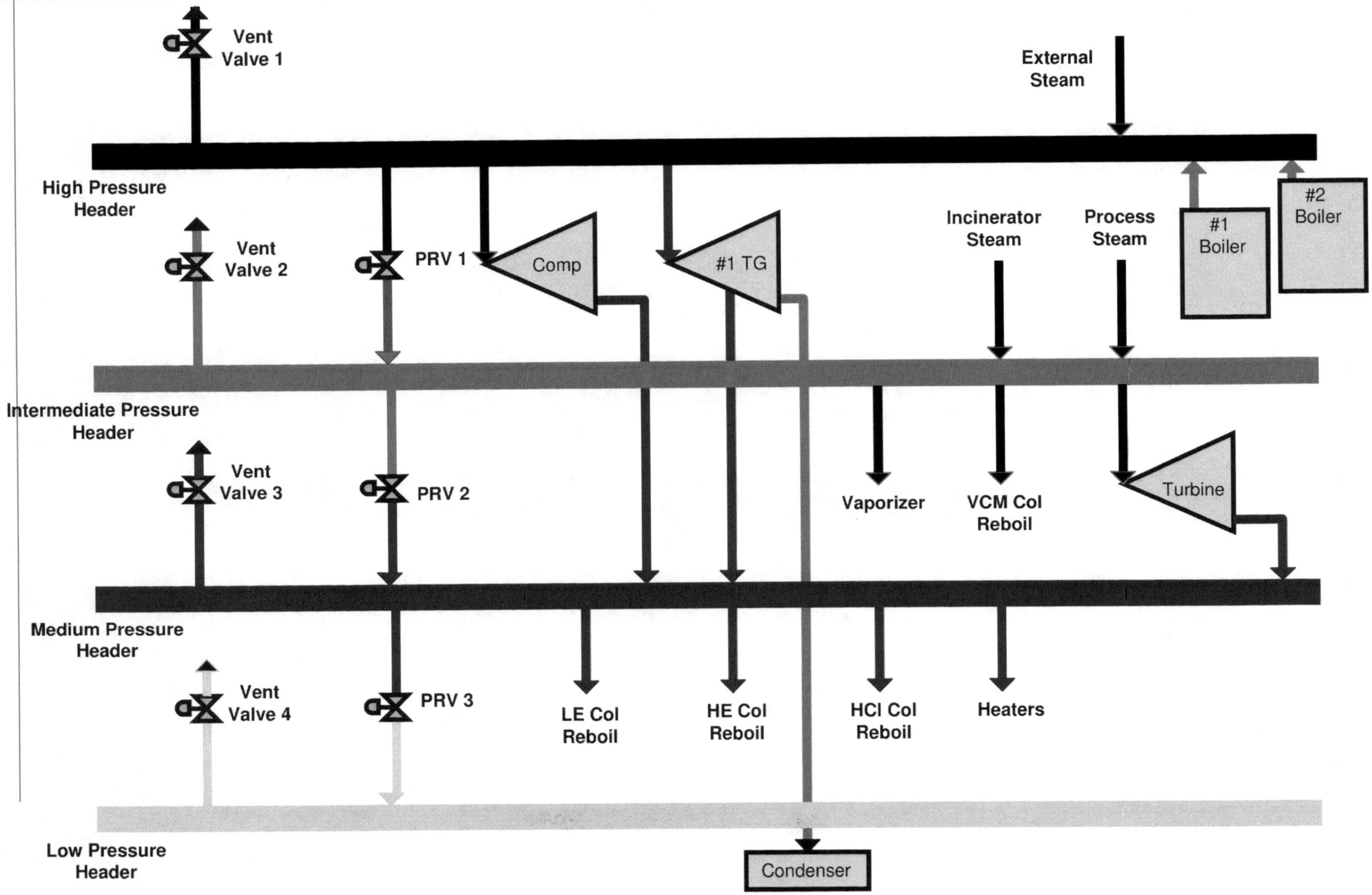

AIChE 2009 Spring National Meeting

Why Do We Care About Process Variability?

Effects of Excessive Variability

- → Product Quality Variability
- → Reduced Equipment Capacity
- → Excessive Energy Consumption
- → Reduced Equipment Reliability and Life
- → Safety and Environmental Concerns

Typical Plant Problems From Excessive Variability

- Chemical - A carbon dioxide plant experienced six shutdowns per month and problems with product quality. The cycle in the output to the ammonia valve upset the refrigeration system.
- Refinery – one third of gasoline blends were off spec initially and require reblending
- Pharma - U.S. pharmaceutical company was unable to start a new plant – could not hold pressure constant in a new centrifuge which meant they could not fill the reagents and washing agents correctly.
- Pulp and paper - mill with three boilers and a steam header
  - Frequent boiler trips for relatively small

Product Quality Variability Effects

- Off – Spec Product
  - Rework, reprocessing, scrap
    - reduced equipment capacity
  - Reduced yields
  - Increased inventory
- Product Quality "Giveaway"
  - Reduced yield
  - Increased energy usage

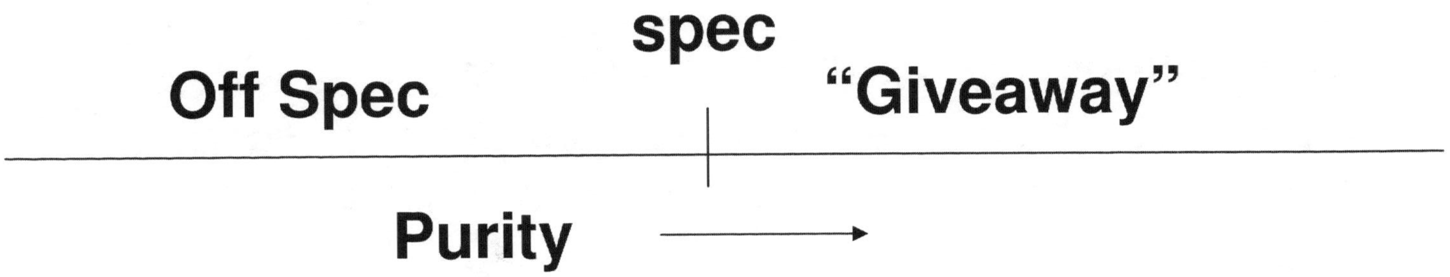

Product Quality Variability Effects Depend (Partially) on Product Blending

- → Products Sold *Without Blending* – Product (often solid) must always be in specification (some can be reprocessed; others scrap)
  - – Paper Machines – other similar sheet manufacturing – textiles, glass, etc.
  - – Batch - incl Pharmaceuticals
- → Products Sold *After Blending* – Final mixed batch must be within specification
  - – Oil Refinery Batch Products – gasoline, diesel, etc.
  - – Major bulk chemicals – fertilizers
  - – Polymers
- → Products with *Limited blending* – Only limited deviation allowed
  - – Pipelined products – natural gas, pure gases (O2, N2, H2), some refinery products
  - – Major intermediary petrochemicals – ethylene, propylene, aromatics, ammonia

Blending Affects Variability Effects!

Product Quality Variability - No Blending Example - Paper Machine

Too thin a sheet leads to a "break" and extra costs.

Scanner for Paper Properties

Final Product

Product Quality Variability

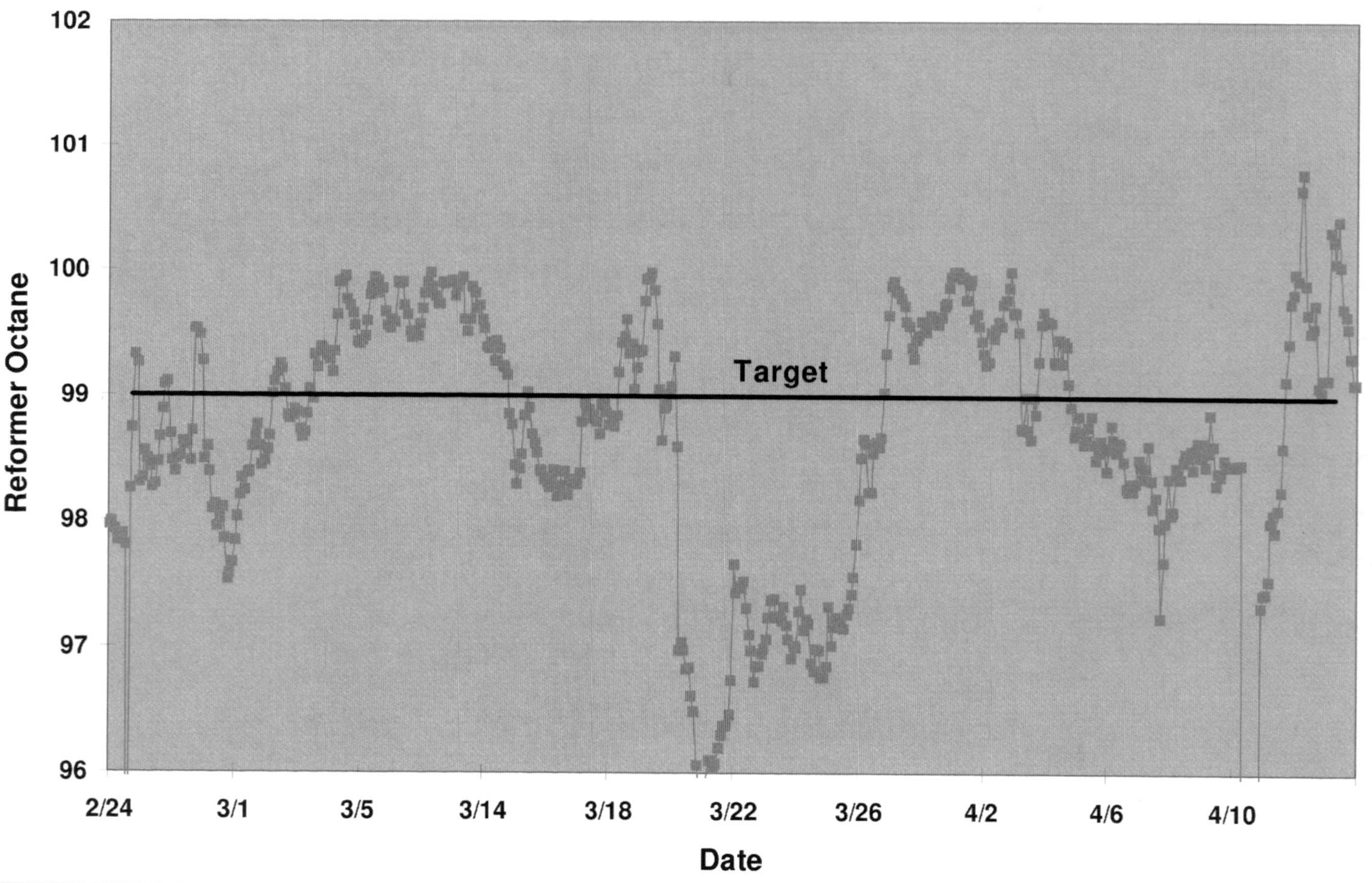

Product Quality Variability - Blended Products

Safety – Draft Control

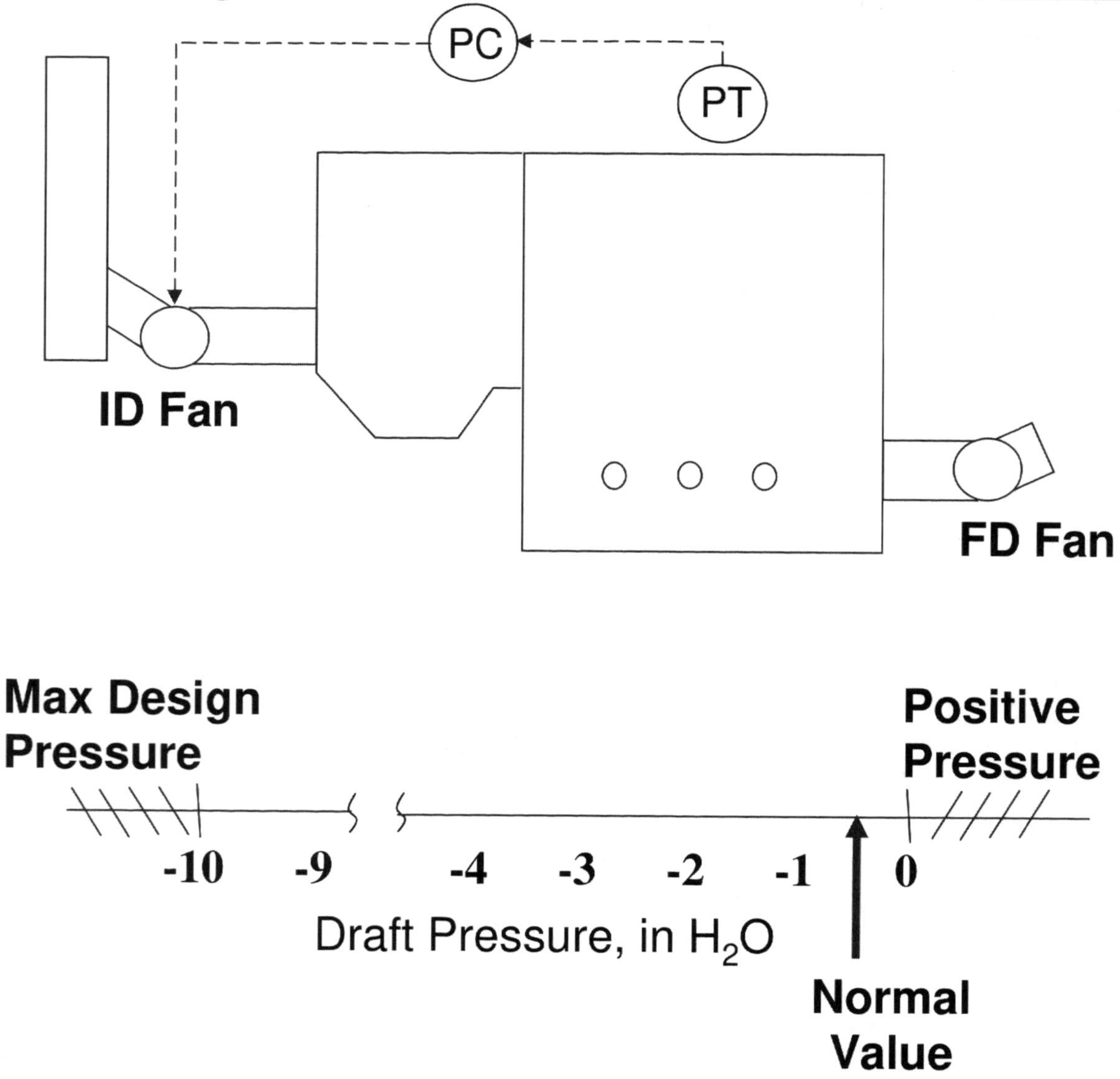

- Boiler draft control is used to remain negative pressure within the boiler.
- If the pressure were to go positive, then hot gases from the boiler could blow back through access ports on operating people

Variability and Safety

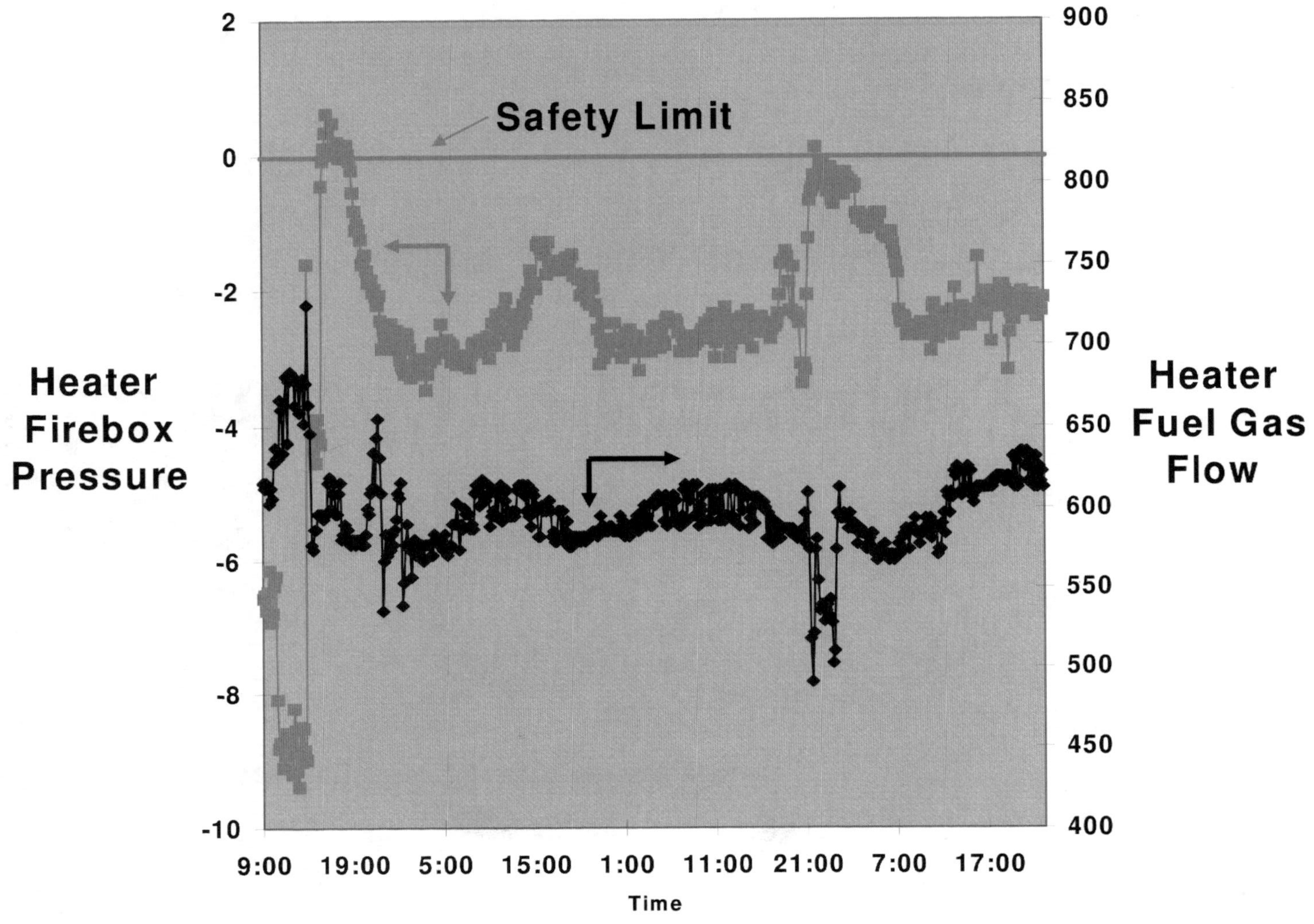

AIChE 2009 Spring National Meeting

How Can We Solve Variability Problems?

Variability

Total Observed Variability =

Disturbance Induced Variability +

Process/ Loop Induced Variability

Reducing Variability

→ ELIMINATE/ REDUCE DISTURBANCES
  – Reduce upstream/ environmental variability

→ Improve Control System Performance
  – Intelligent System (not just Loop) Tuning
  – Improved Valve Performance
  – Better, Faster Measurements
    • Reduce Time Delay in Loop
  – Compensate for Process/ Loop Non-linearities
  – Control Loop(s) Reconfiguration
  – Advanced Control Technology

→ Process Modifications

→ Increase Product Blending (If Possible)

Process/ Loop Induced Variability - Potential Problems

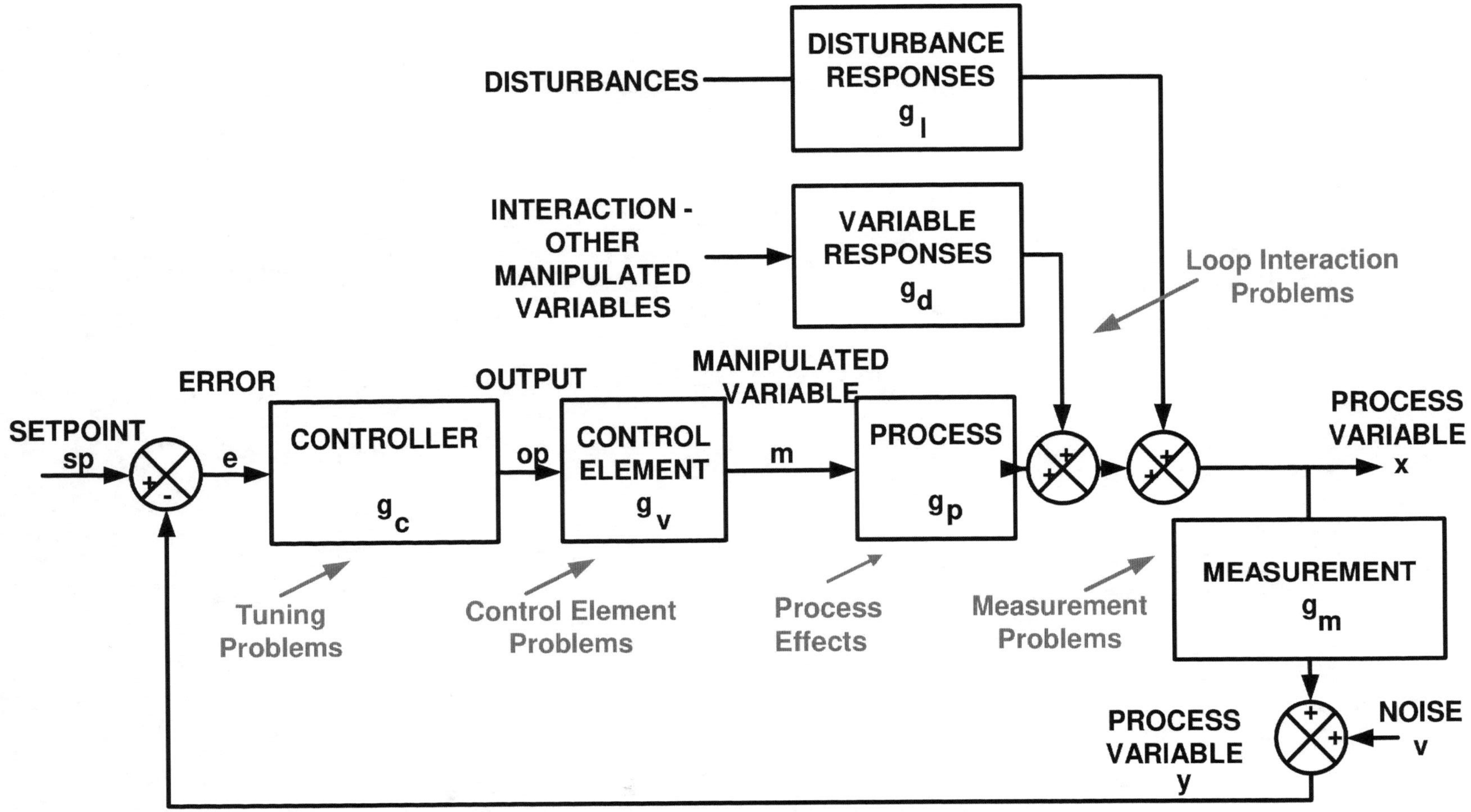

AIChE 2009 Spring National Meeting

Types of Process Responses

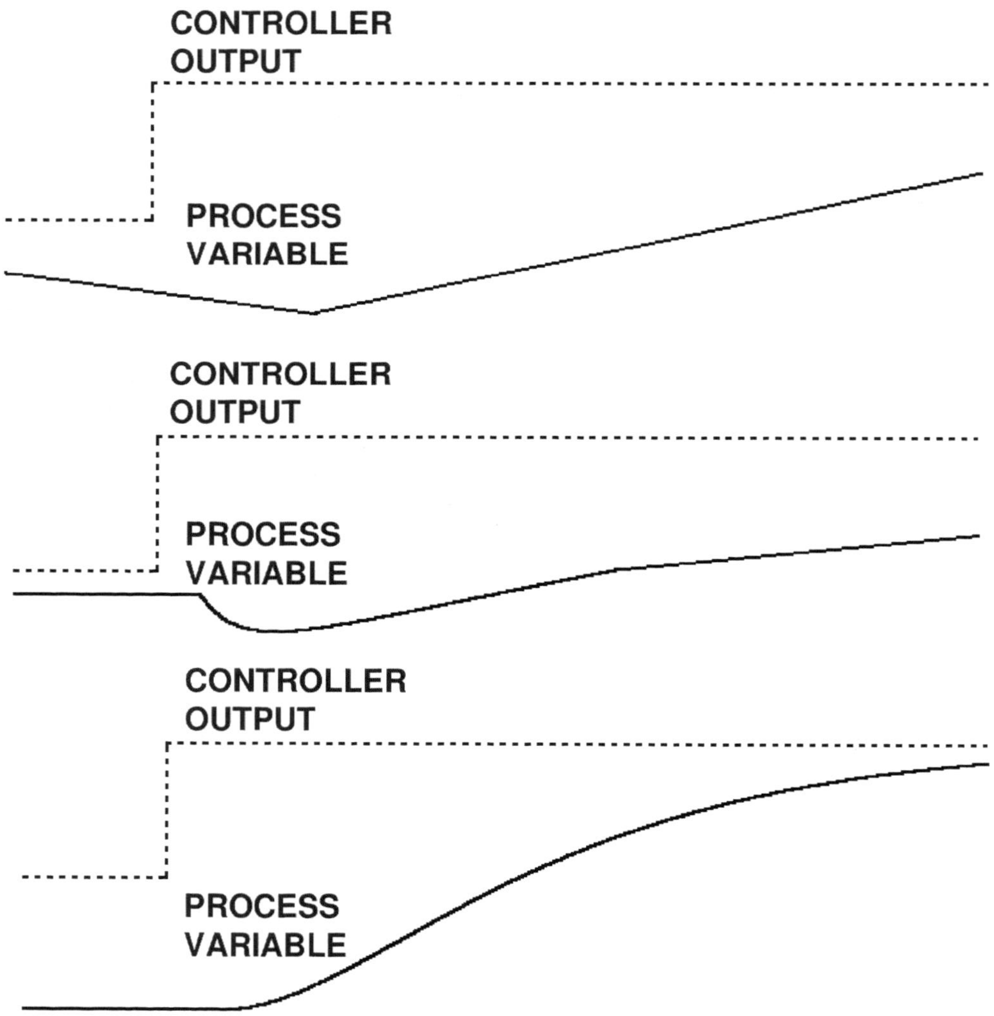

Level
(Integrating)

Boiler Drum Level
Inverse Response

Temperature
Self- Regulating

Cyclic Behavior – Limit Cycles

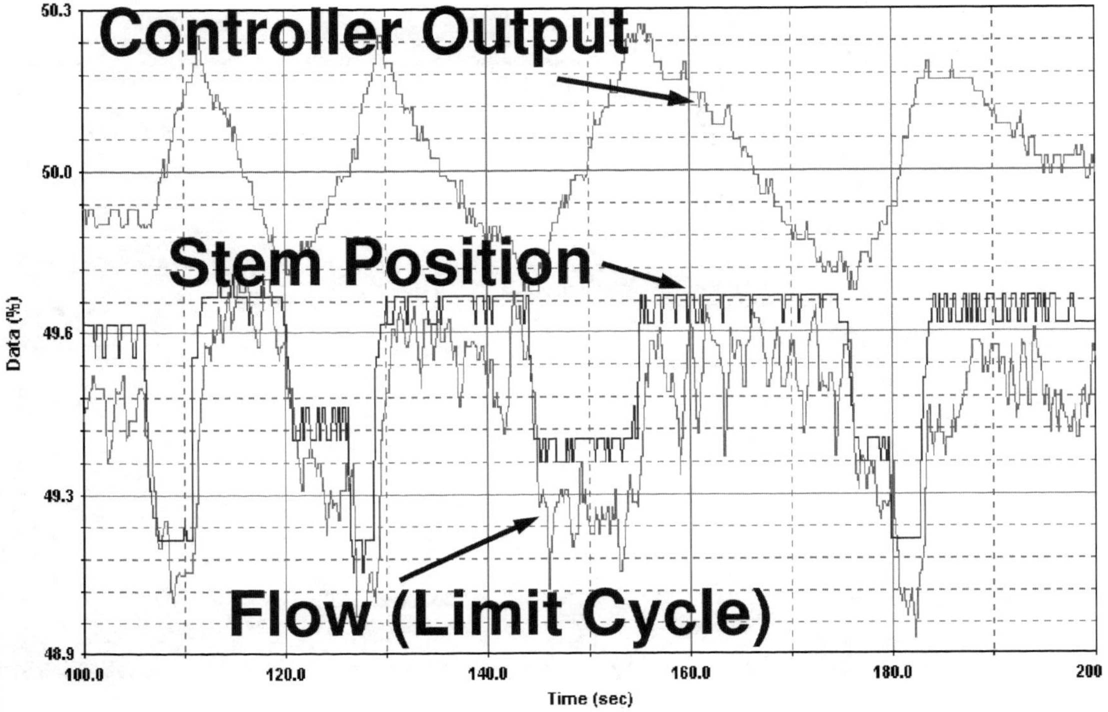

Integrating Process (Level, Pressure (Some), Others)

+ Valve Stiction

= Limit Cycle

Solution:

- **Reduce Stiction**
- **Proper Level Loop Tuning**

Cyclic Behavior – Valve Problems

Limit Cycle in Auto

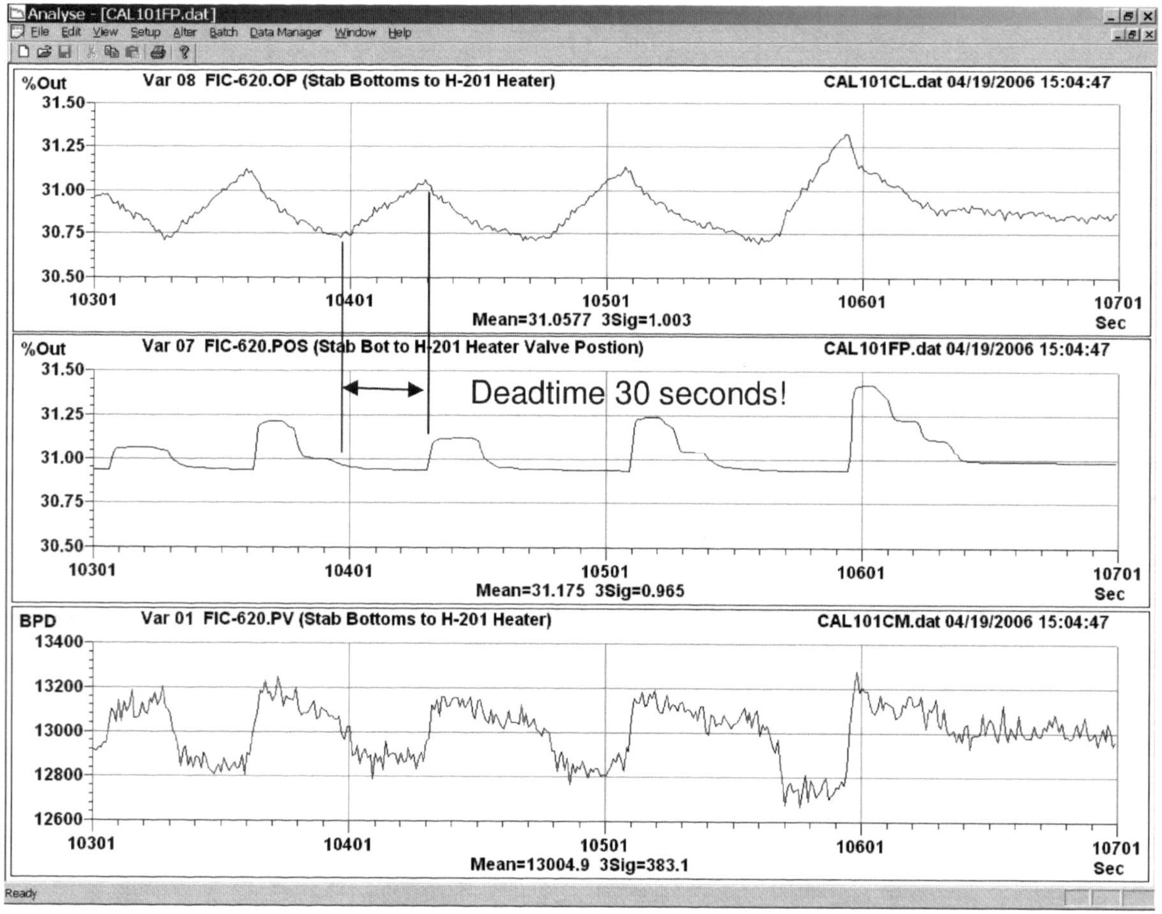

Controller Output

Actual Valve Position

Process Variable

AIChE 2009 Spring National Meeting

Cyclic Behavior – After Valve Problems Fixed

~ No Limit Cycle in Auto

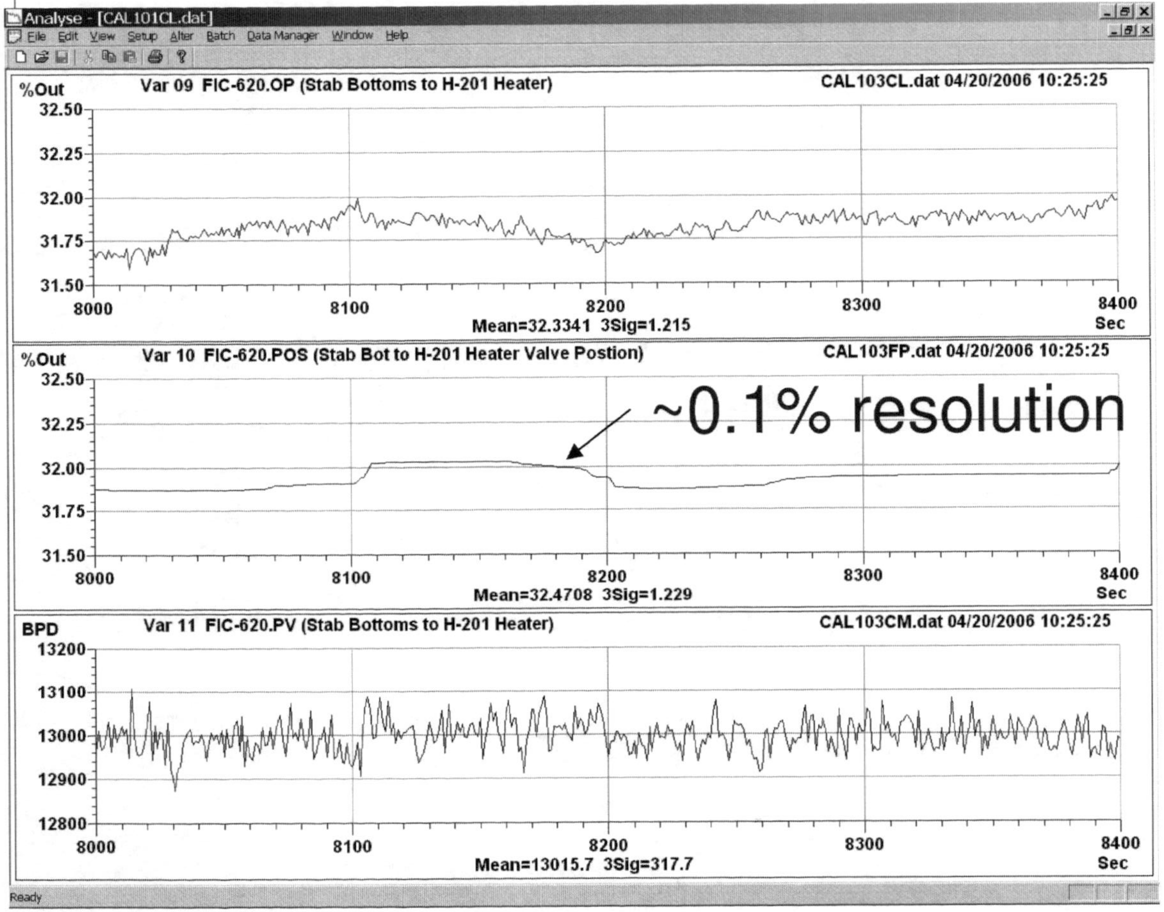

Controller Output

Actual Valve Position

Process Variable

Cyclic Behavior – Tuning Example

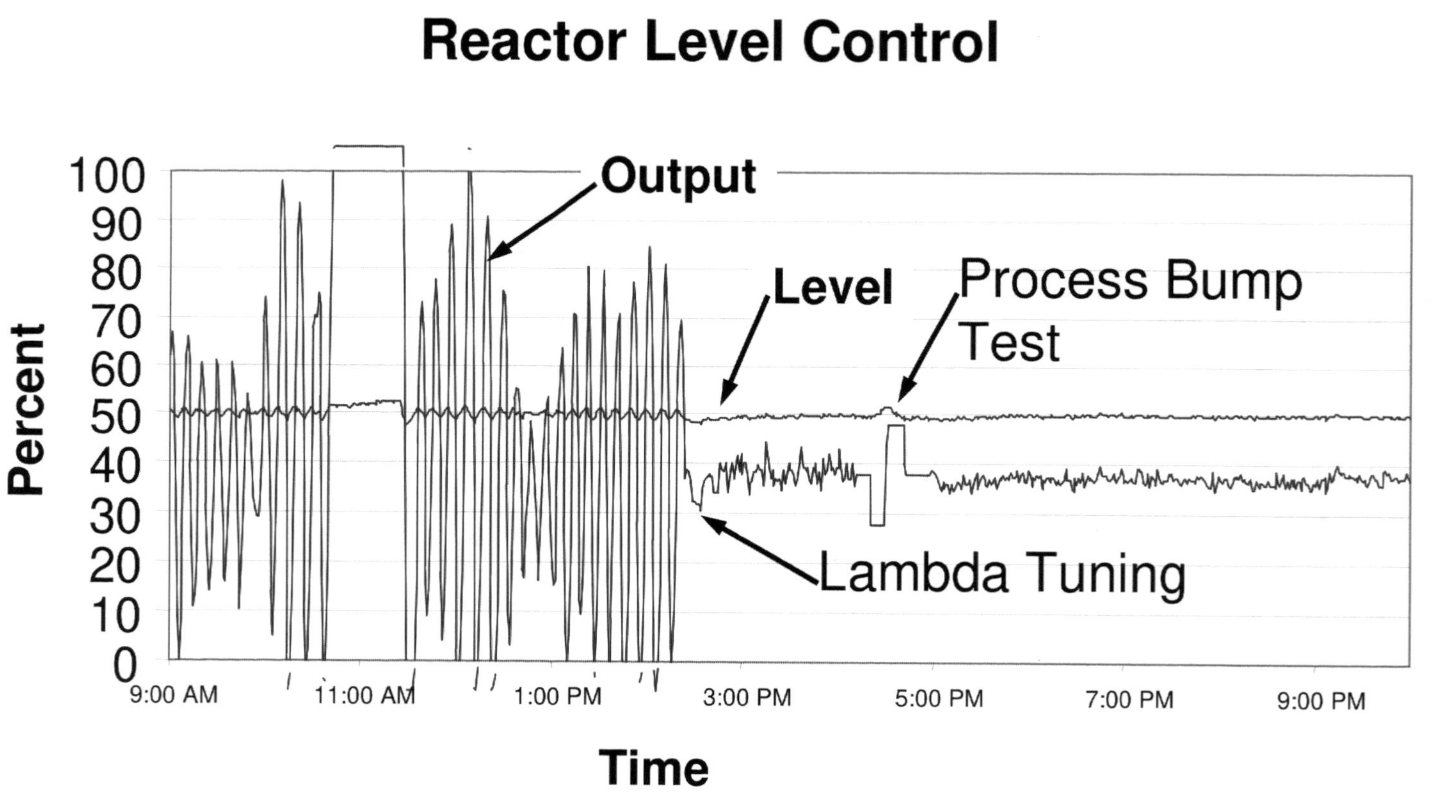

Original Tuning 6.0 / 0.25 min/rep
New Tuning: 4.68 / 41.5

How does excessive variability affect plant economics?

Typical Process Plant Operation

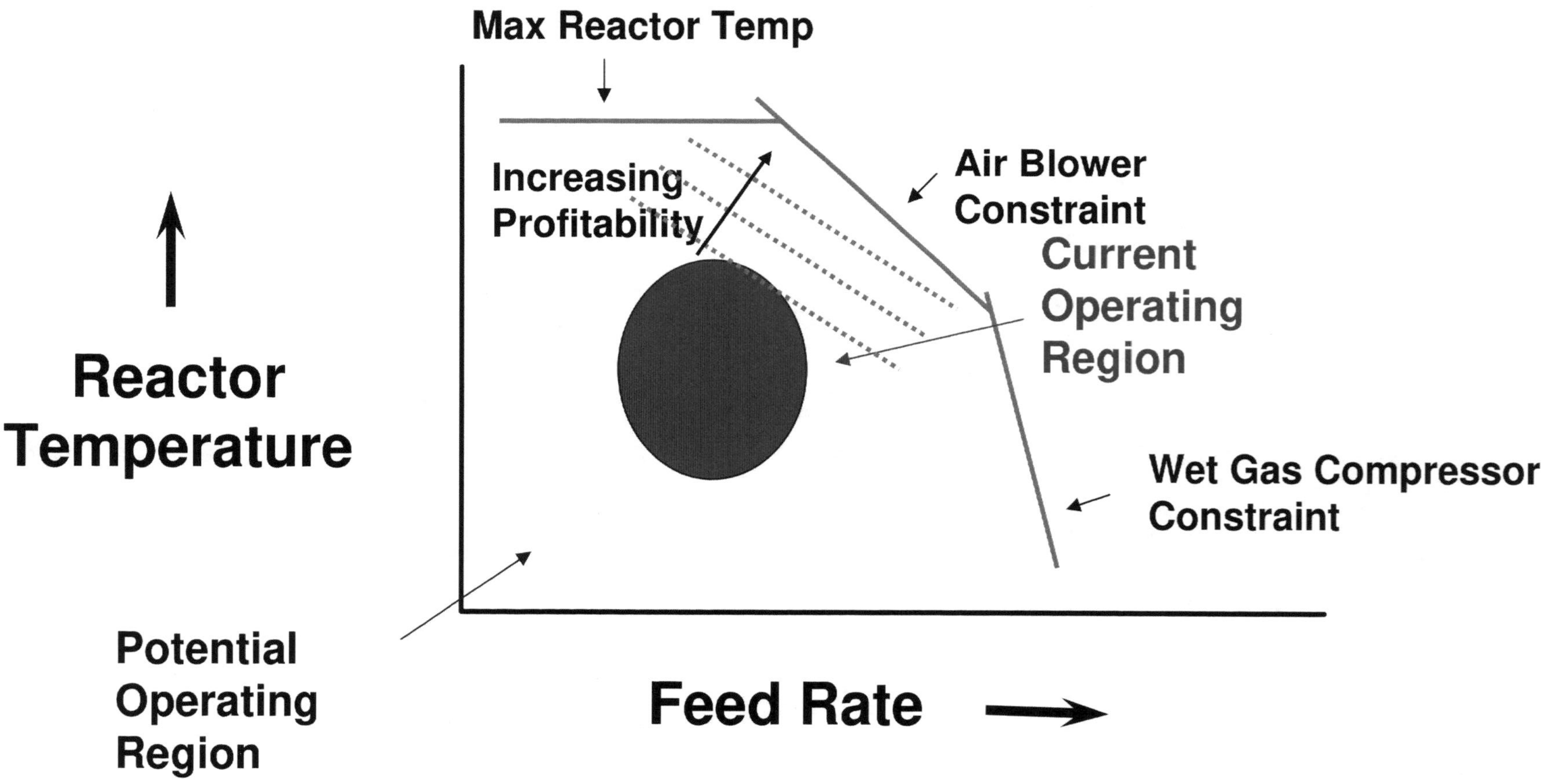

Typical Process Plant Operation

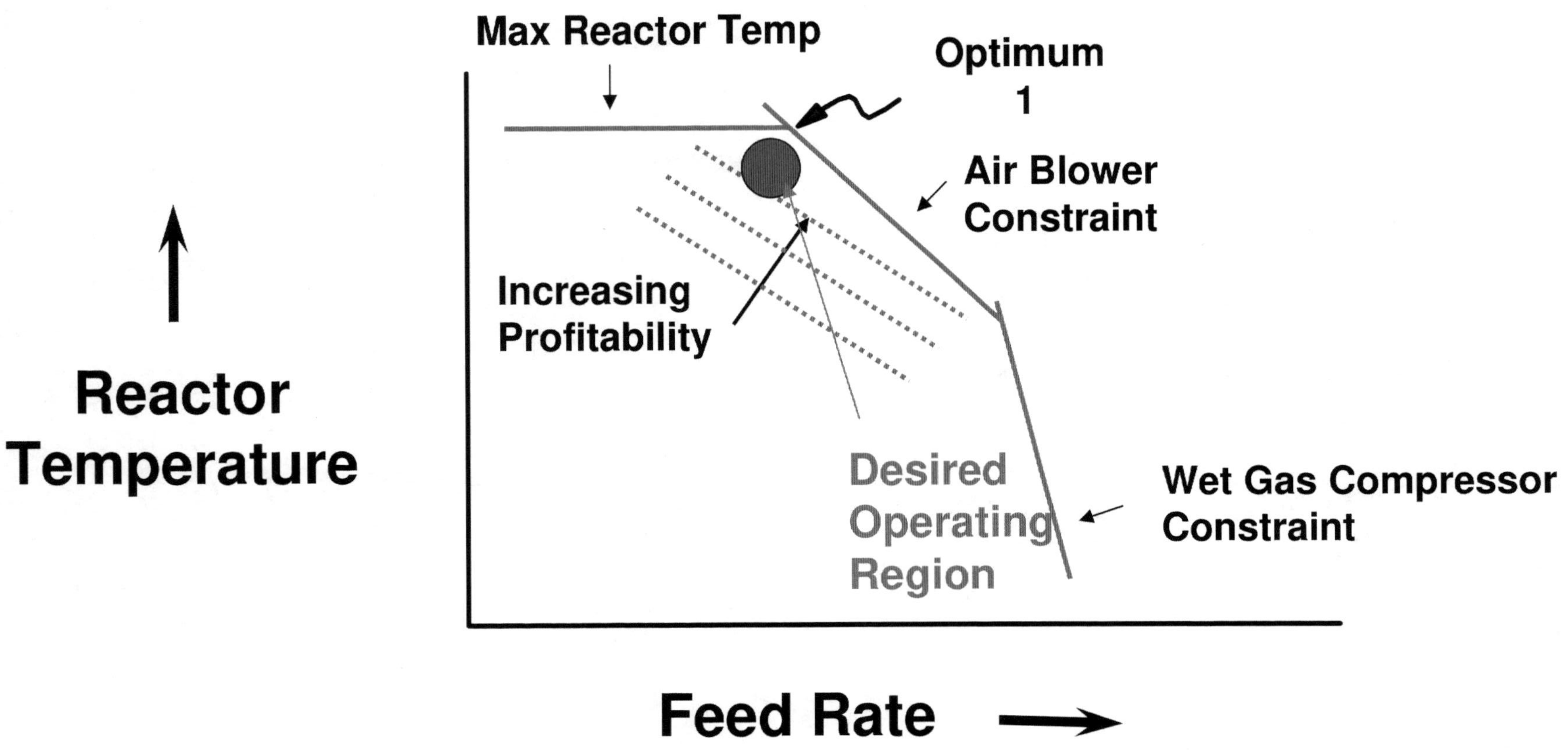

Representation of Variability

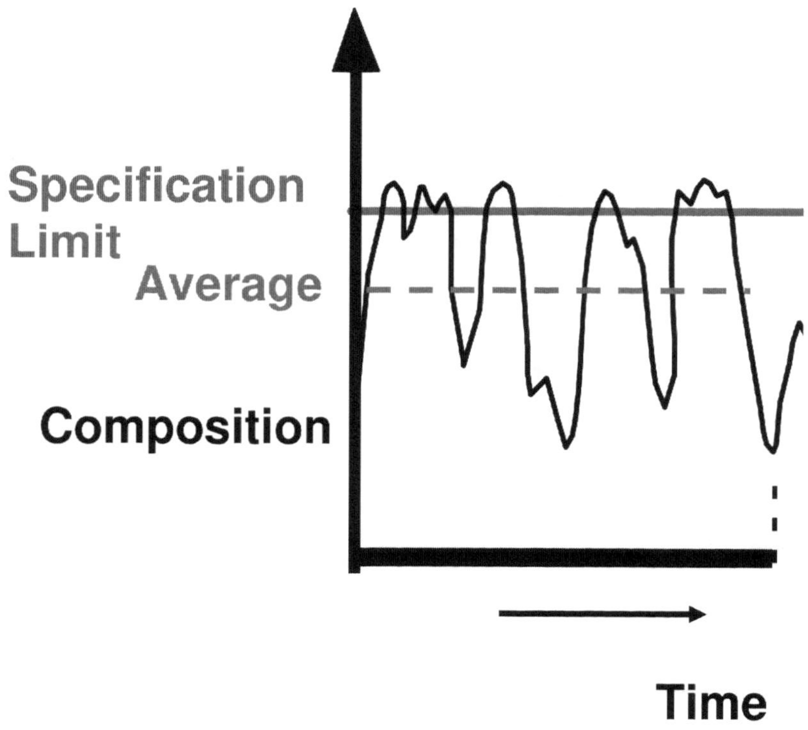

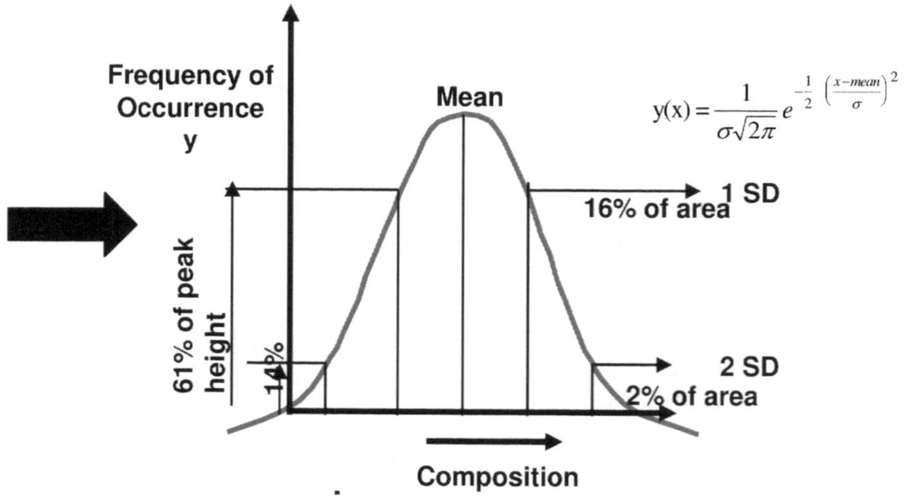

AIChE 2009 Spring National Meeting

Effect of Reduced Variability

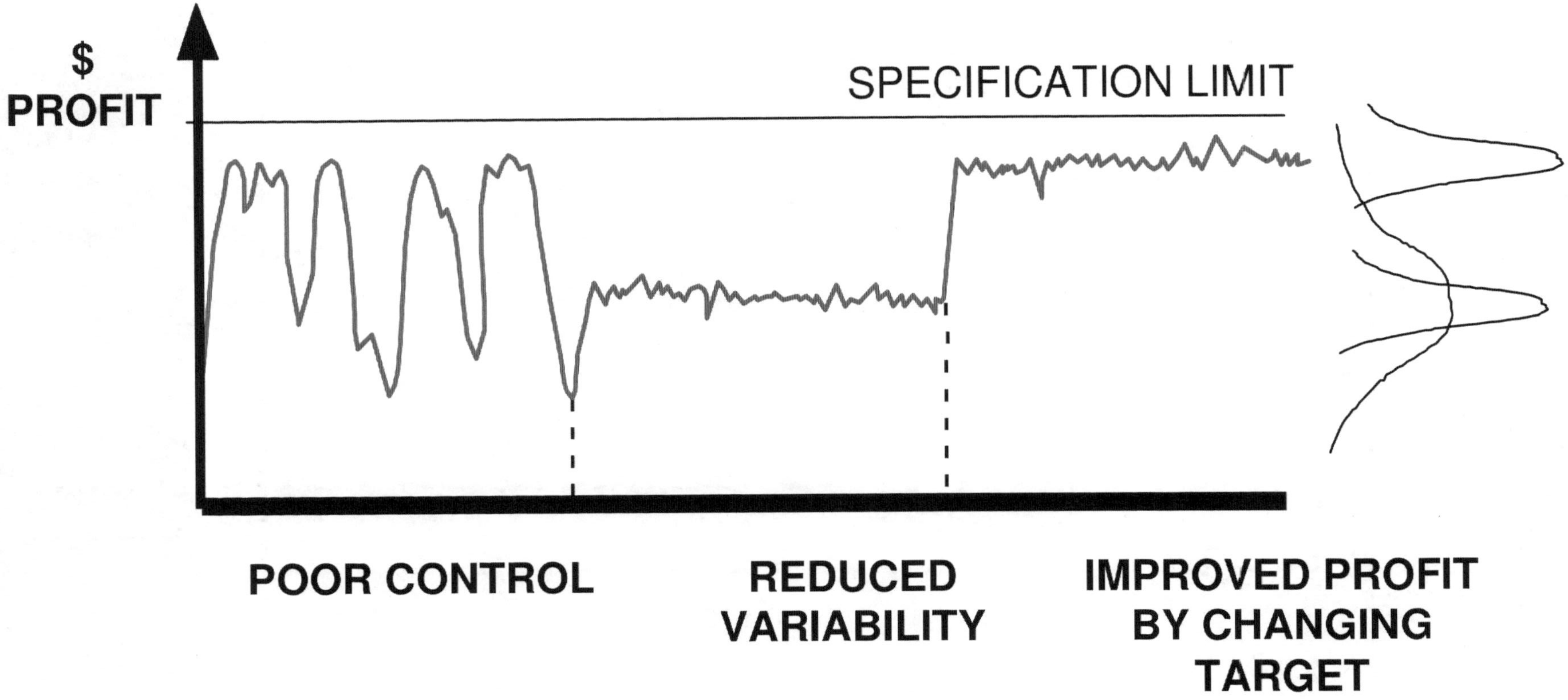

Effect of Variability – Linear Objective Function

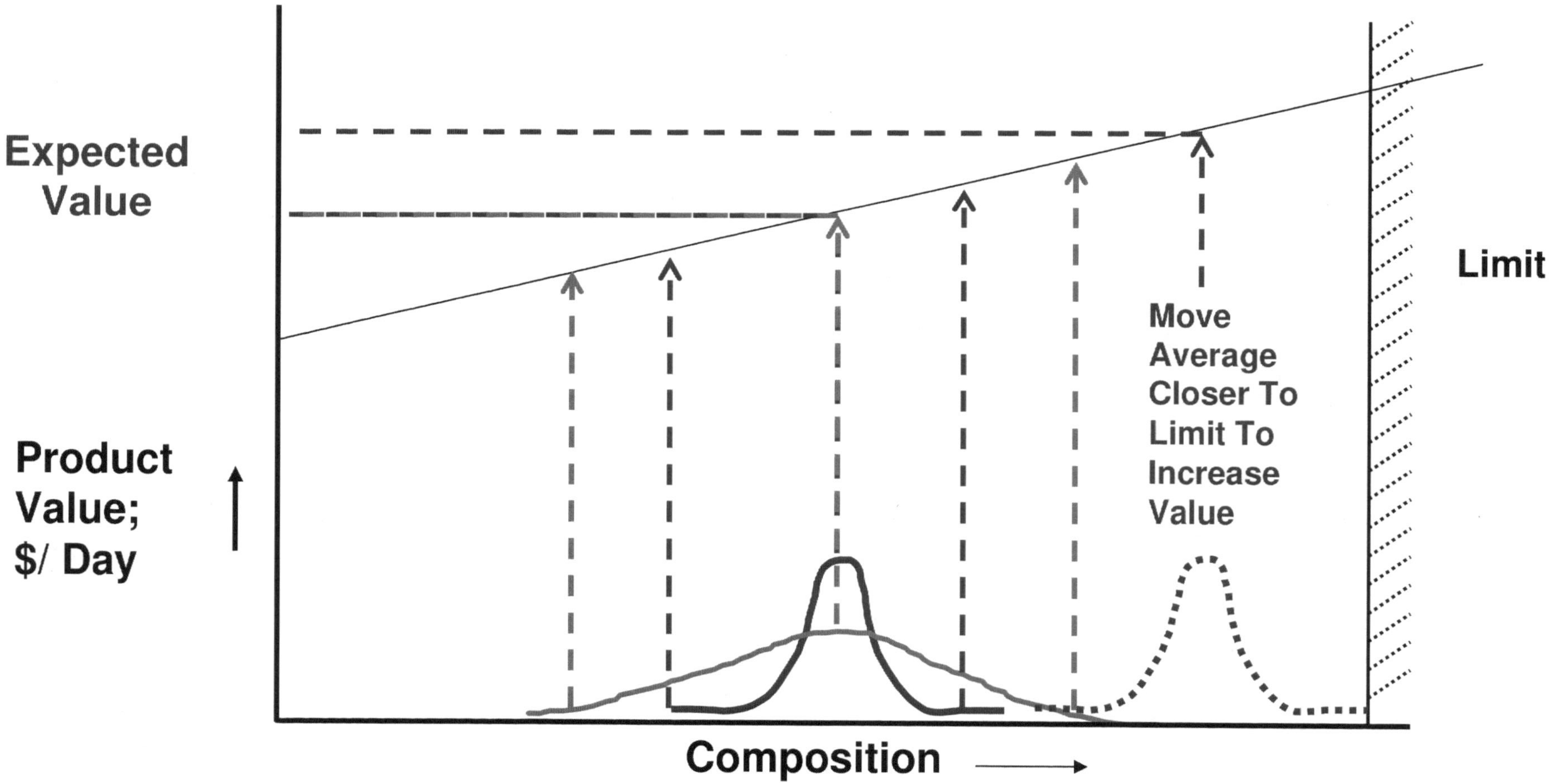

No Change in Expected Value With Change in Variability- only with Change in Mean!

AIChE 2009 Spring National Meeting

Economic Value of Variability Reduction

- → **Increased Production (Production Limited - Plant at maximum capacity)**
  - Increased Equipment Productivity
  - Improved Equipment Reliability
- → **Constant Production (Market Limited)**
  - Increased Yield Most Valuable Product (Reduced raw material costs)
  - Reduced Operating Costs
    - Energy
    - Maintenance
    - Shutdown frequency

Calculating Economic Benefits – Linear Objective Function

- → Calculate Base Case Variability
- → Estimate Potential Reduction In Variability From Enhanced Automation
- → Project Change In Average Value For The Operating Variable With Reduced Variability
- → Calculate Change In The Economic Objective Function Corresponding To The Change In The Average

Case Studies

Fluid Catalytic Cracking Unit

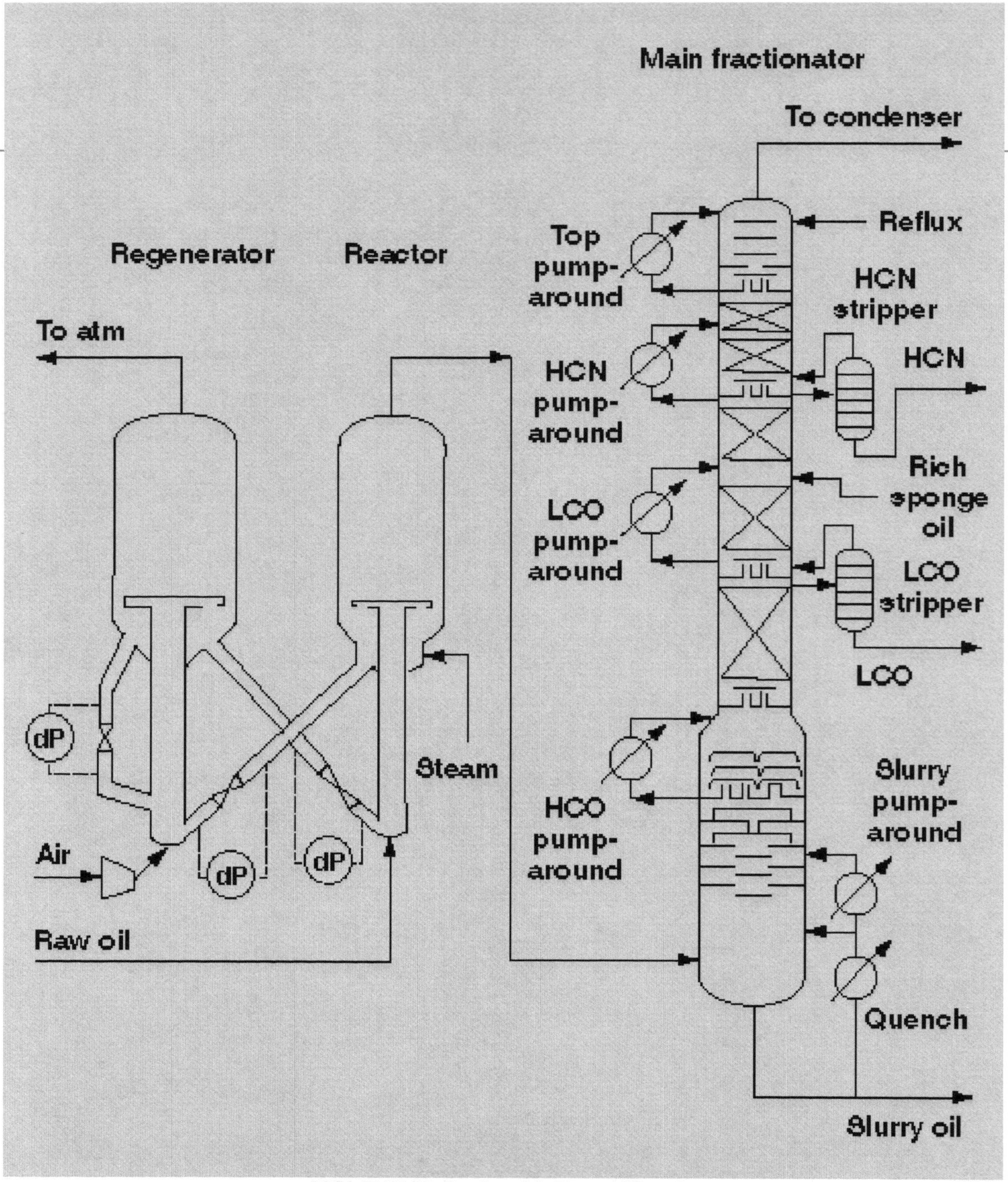

Case Study – Calculation Variability Benefits – Increased Production - Feed Rate Statistics

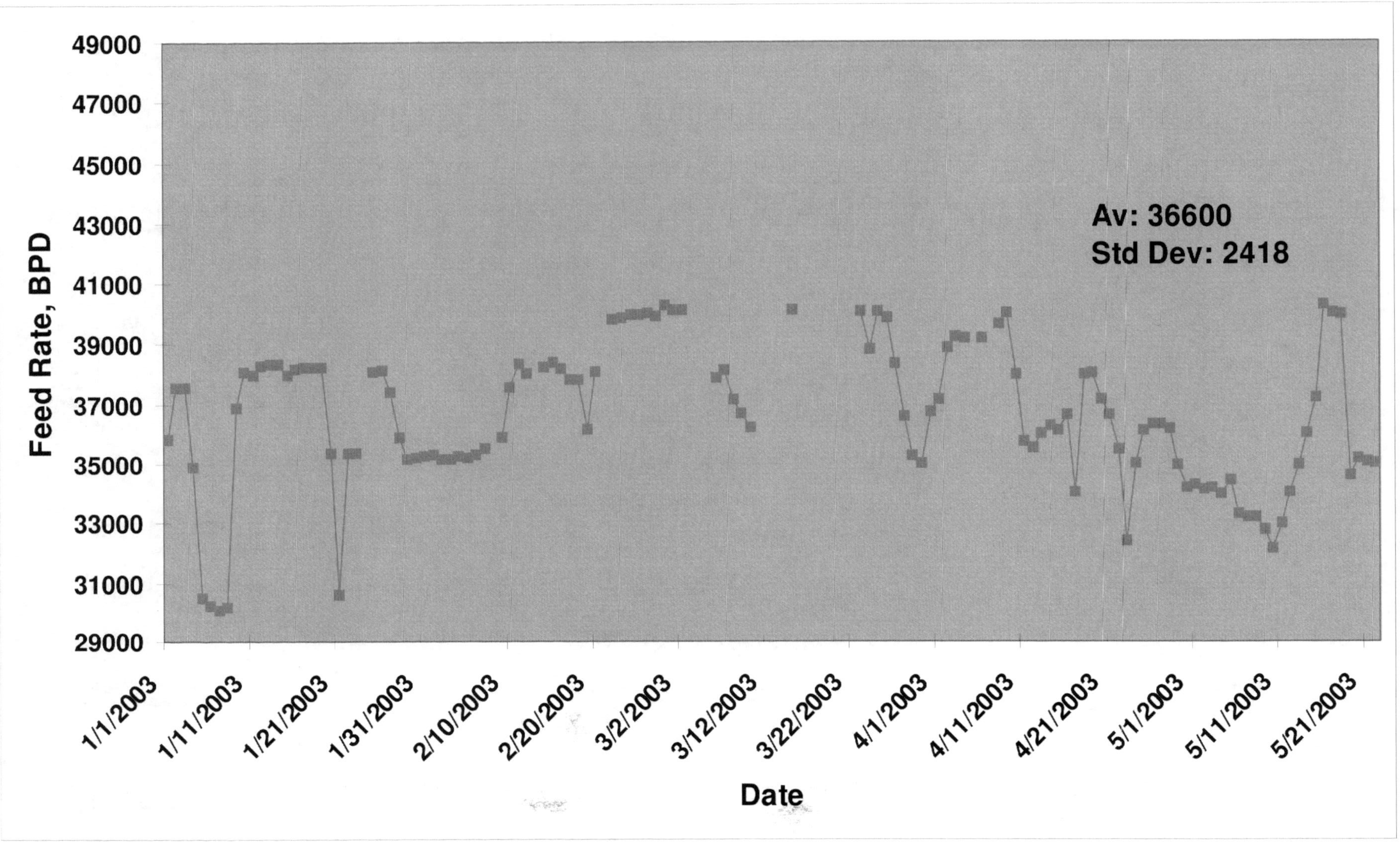

Case Study – Calculation Variability Benefits – Increased Production – Constraint Statistics

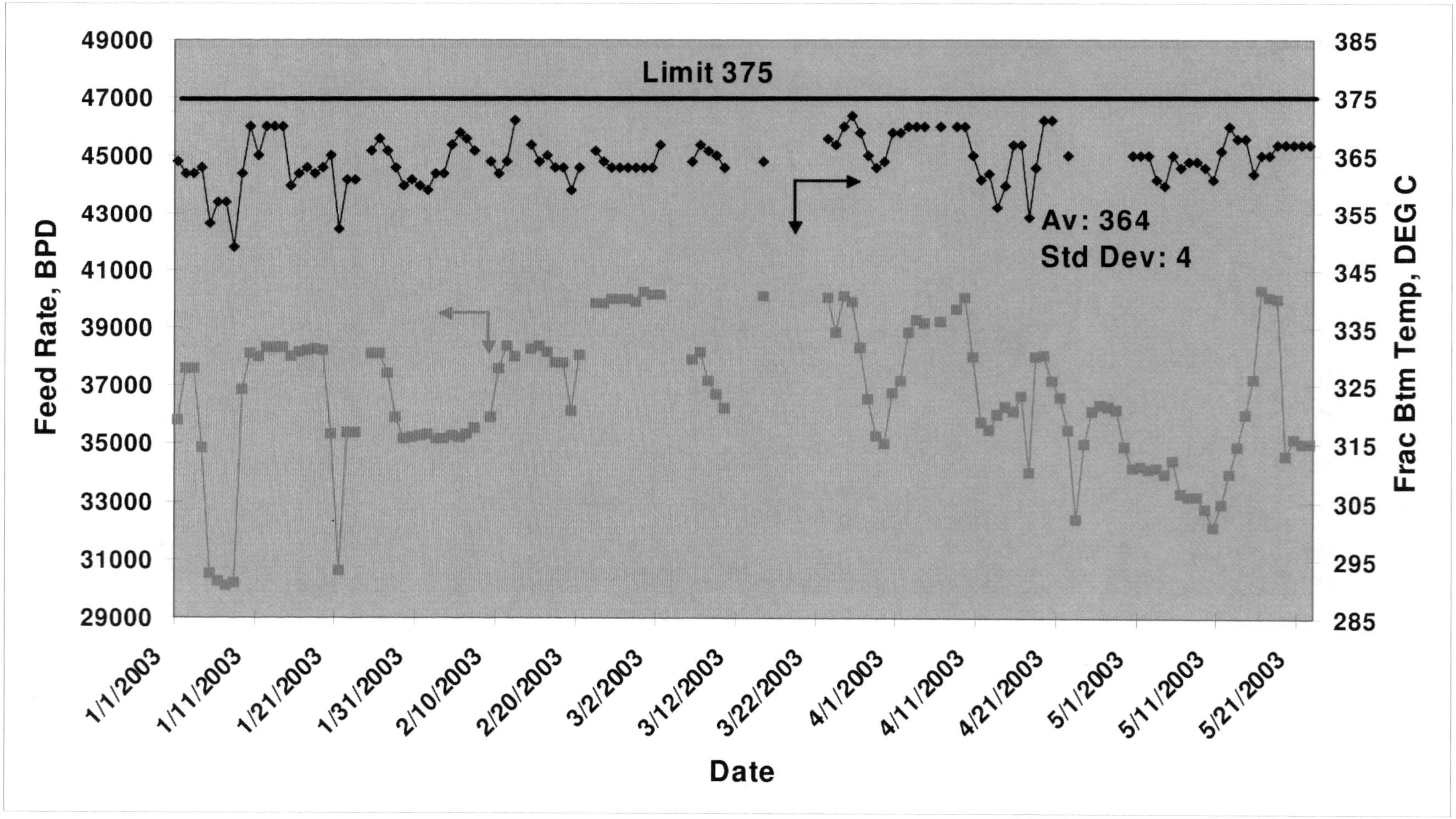

AIChE 2009 Spring National Meeting

Case Study – Calculation Variability Benefits – Increased Production – Constraint Statistics

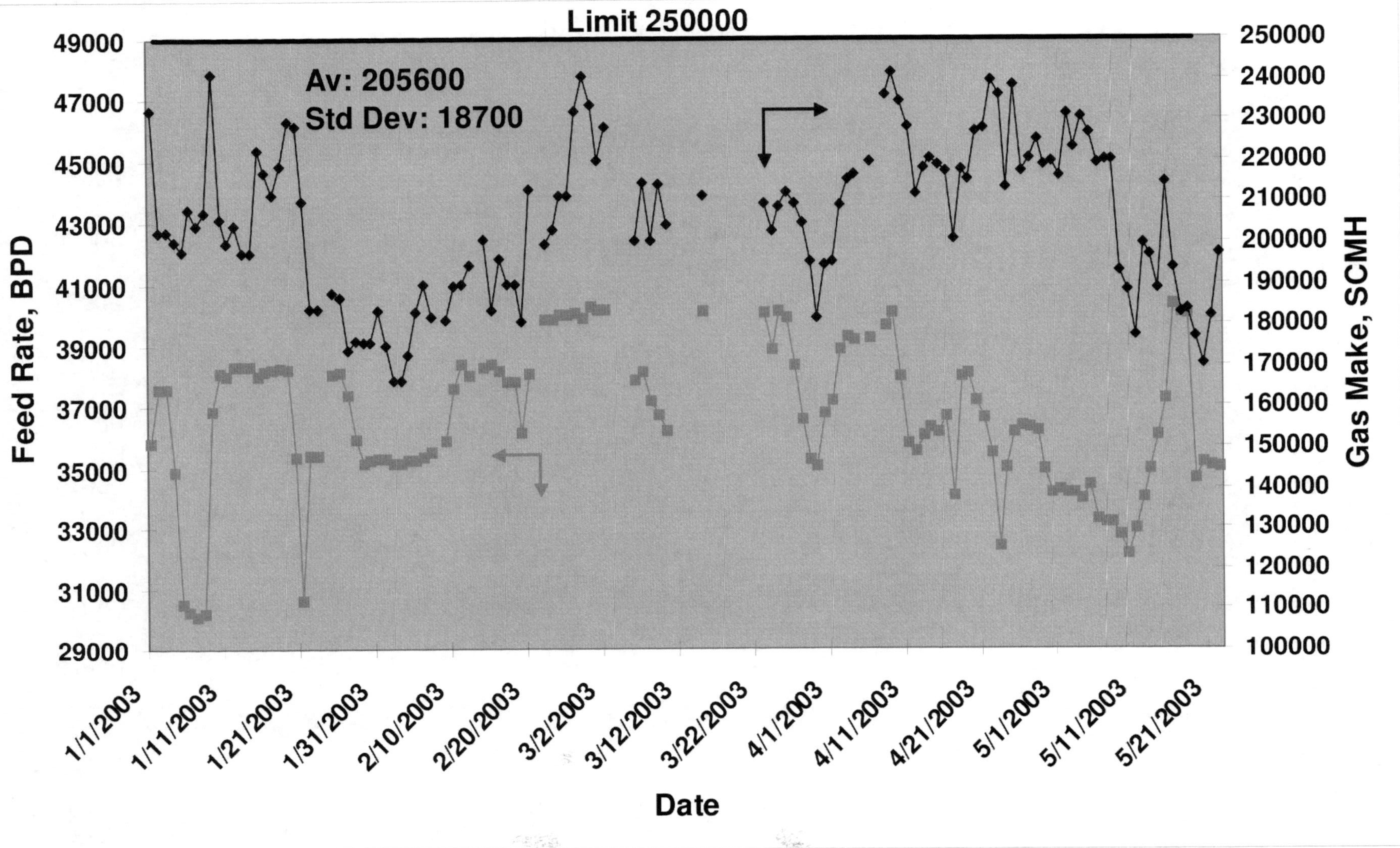

AIChE 2009 Spring National Meeting

Case Study – Calculation Variability Benefits – Increased Production – Constraint Statistics

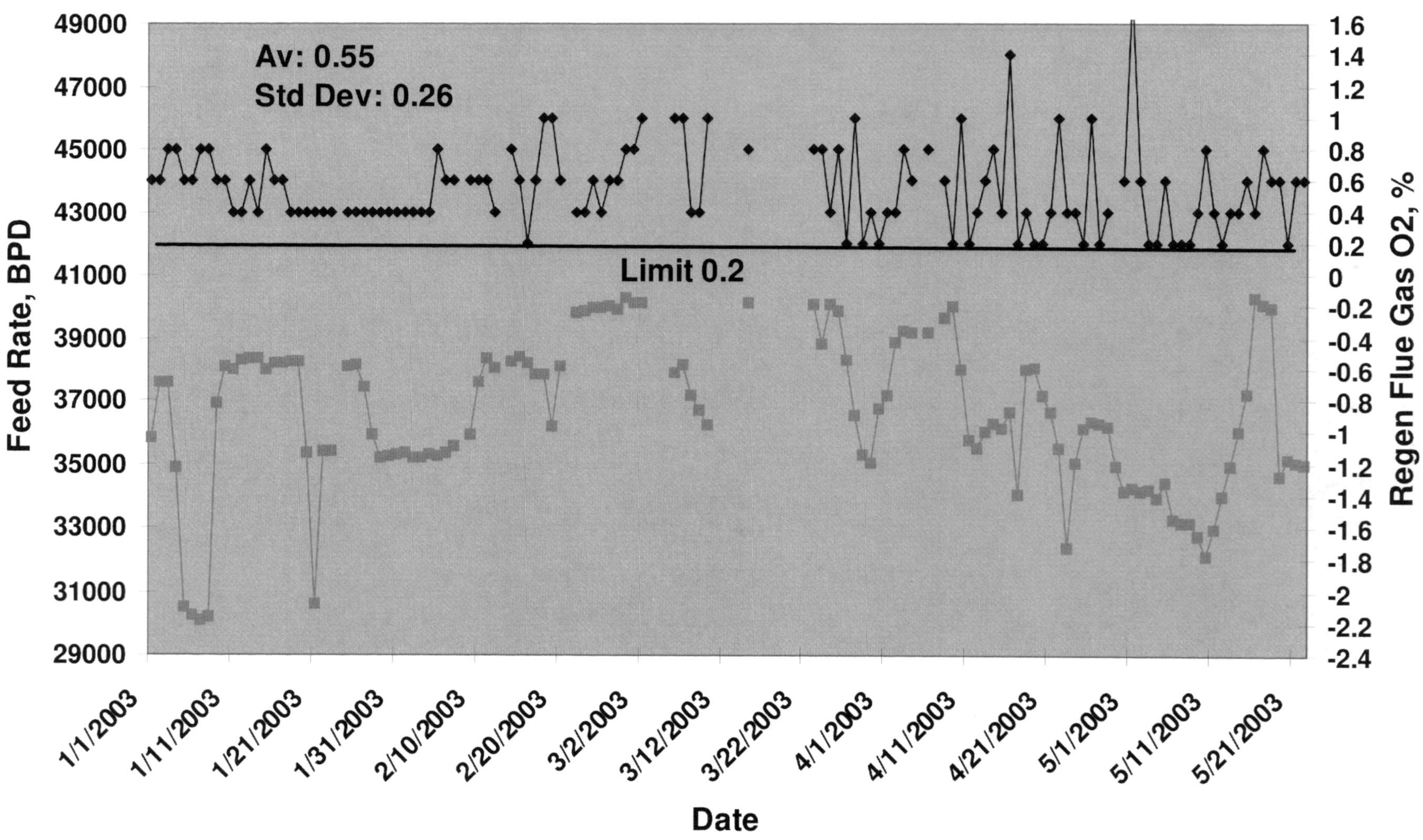

Case Study – Calculation Variability Benefits – Increased Production – Constraint Statistics

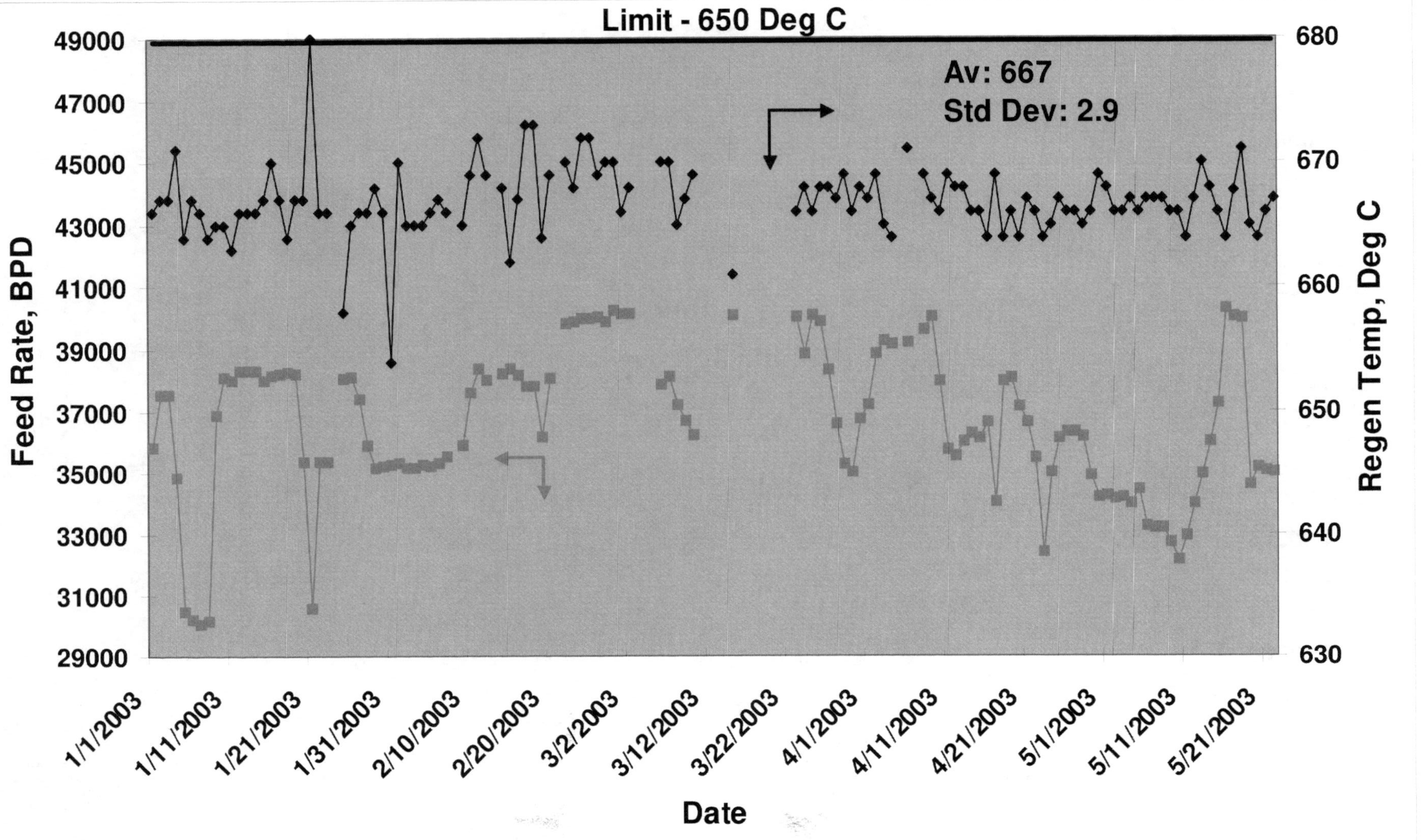

Case Study – Calculation Variability Benefits – Market Limited Operation

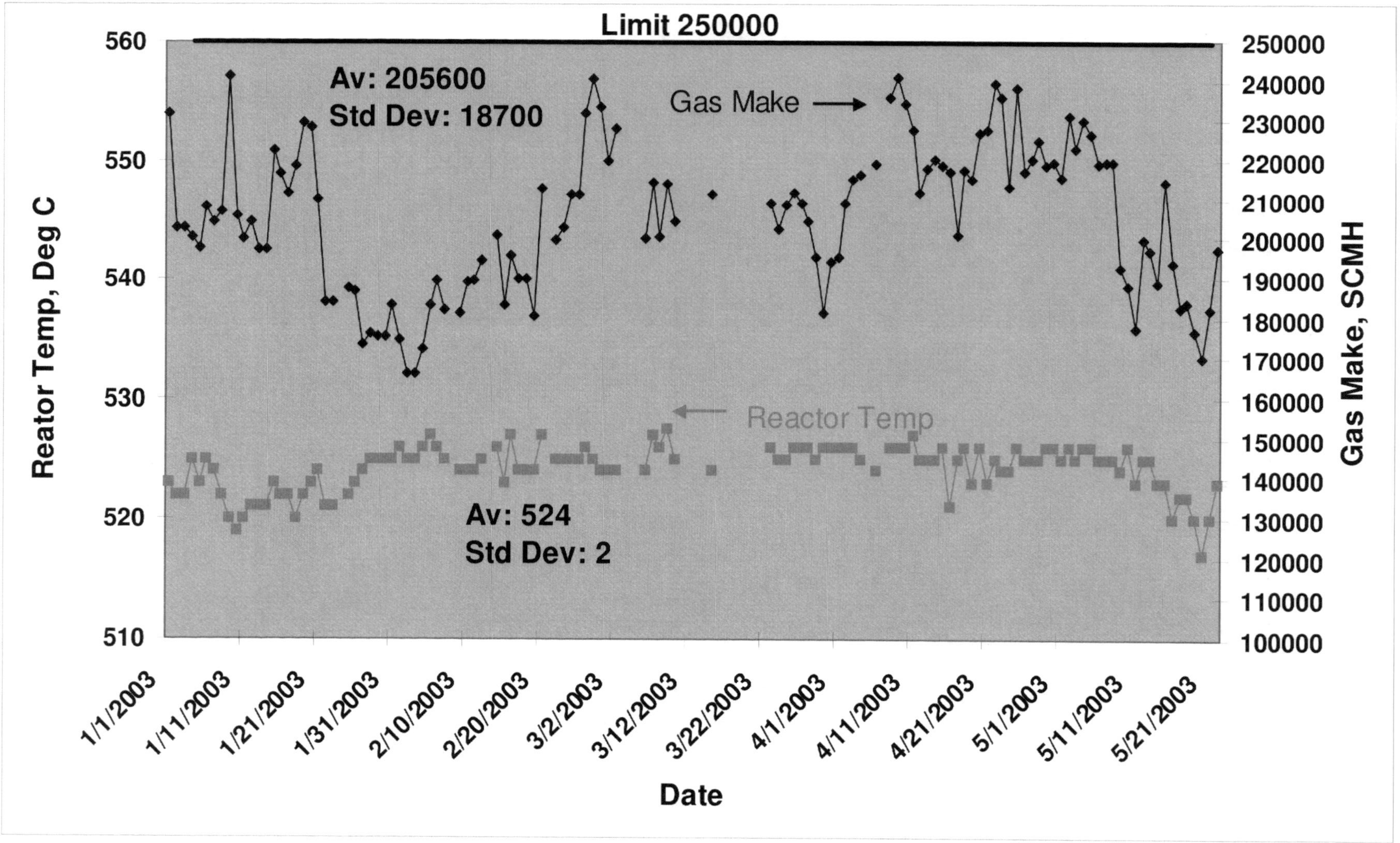

Case Study – Calculation Variability Benefits – Product Quality

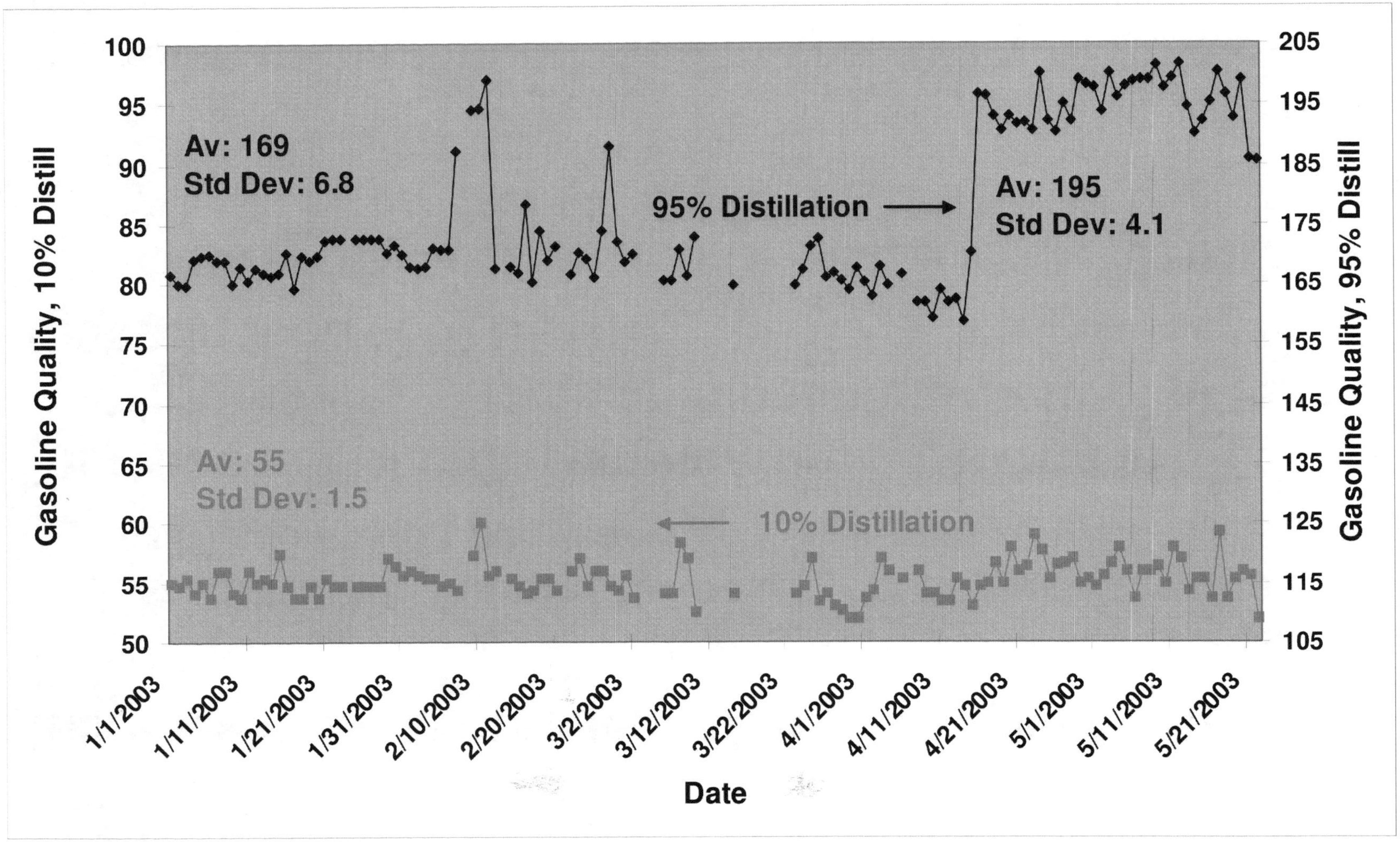

AIChE 2009 Spring National Meeting

Potential Benefits from Variability Reduction

- Increase in maximum feed to unit of at least 3% (1200 BPD)
  - At US$10/ Bbl margin; 200 days/ year at maximum rate - US$2.66 Million per year
- Increase in Reactor Temperature when not at maximum rate – Increase in gasoline yield of at least 0.5% on feed (240 BPD)
  - At US $8/ Bbl total incremental product value; 100 days per year - US$ per year - $192,000
- Increase in maximum gasoline recovery of at least 0.5% on feed (240 BPD)
  - At US $6/ Bbl margin of gasoline over LCO – US$ 480,000 per year

Case Studies

Ammonia Process Example

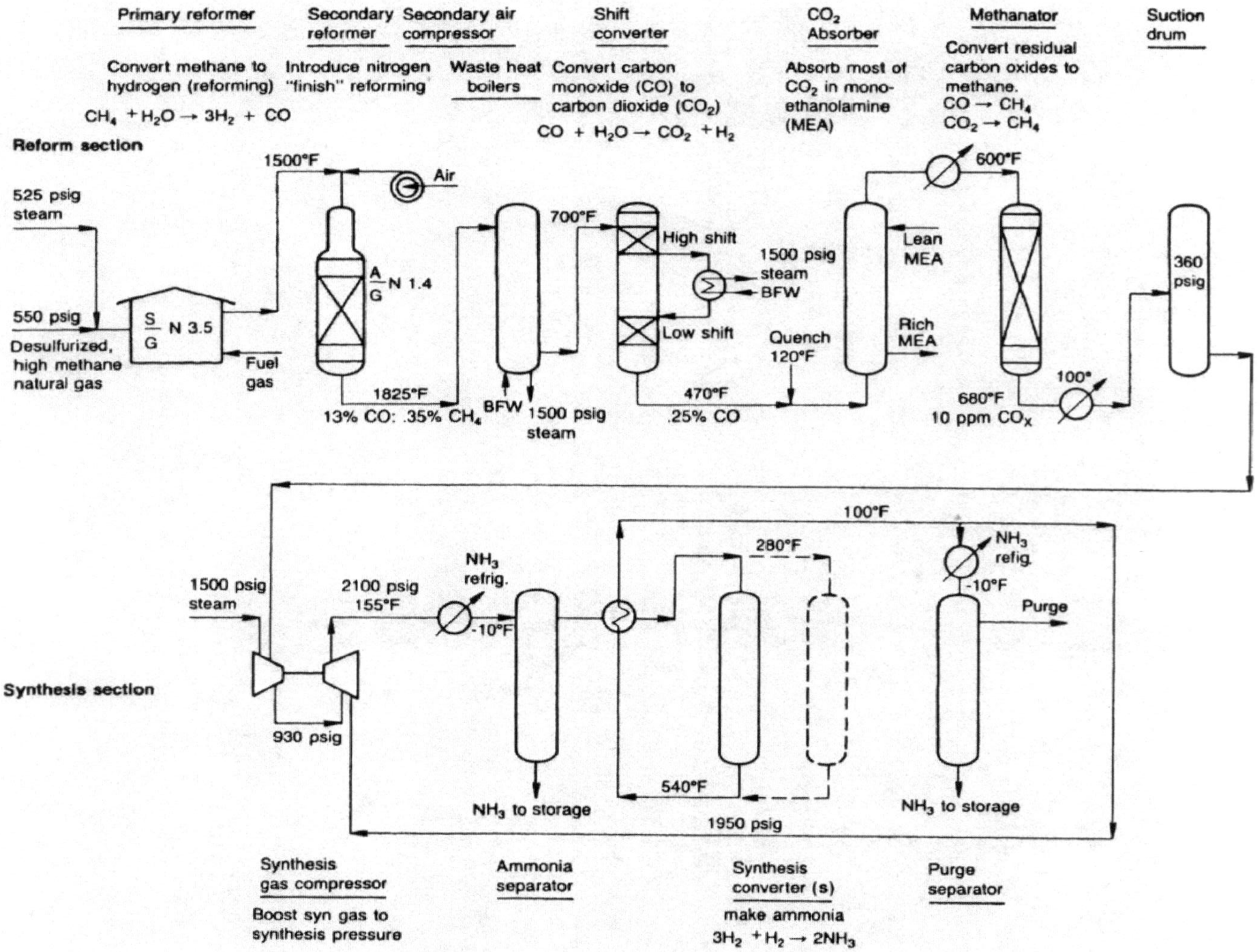

Synthesis Converter Bed Temperature

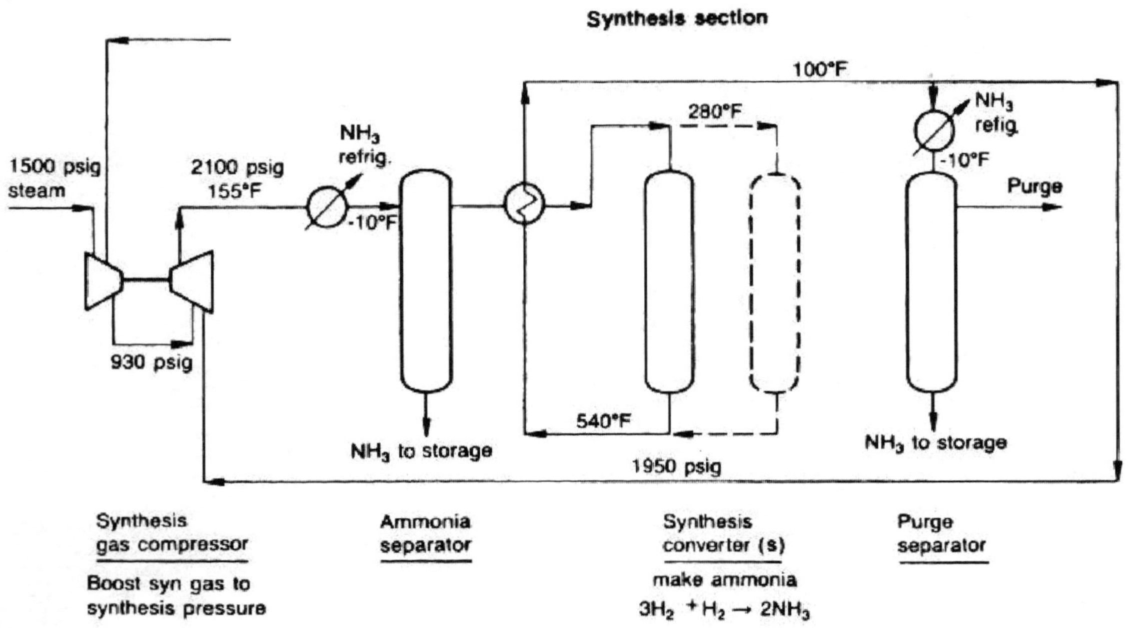

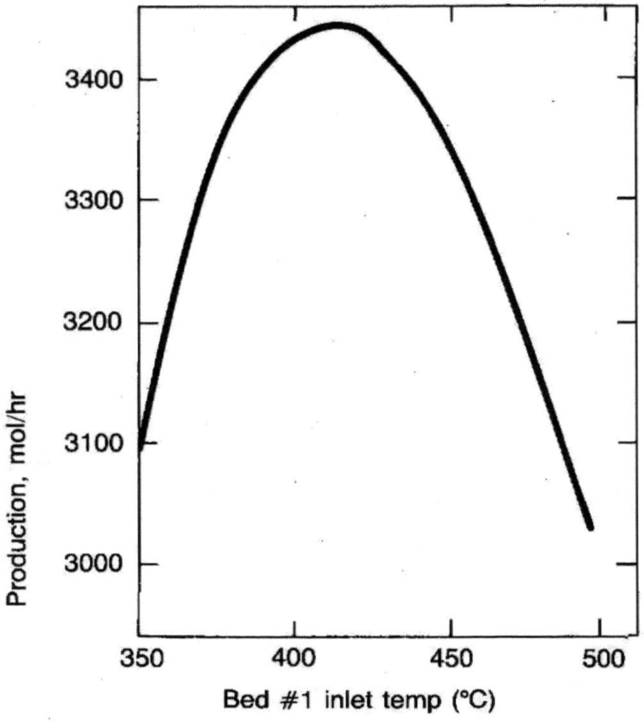

AIChE 2009 Spring National Meeting

Impact of Synthesis Bed Temperature on Ammonia Production

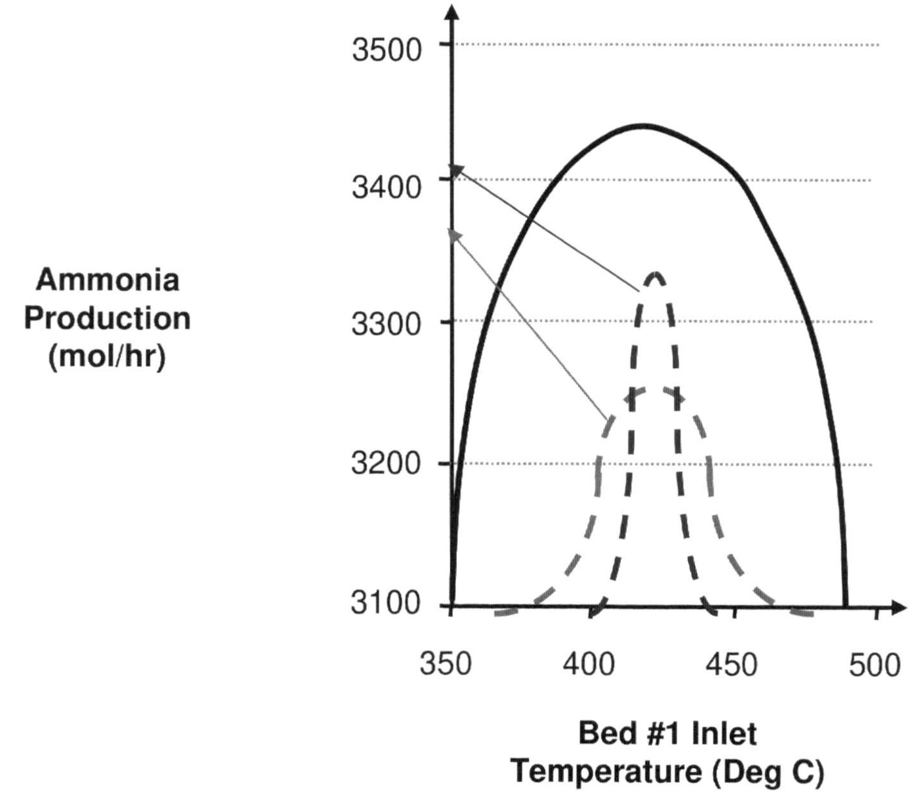

Mean (DegC)	Variation (DegC)	Production Loss(%)
415	0	0
415	25	0.103
415	50	0.415
460	25	1.102
460	50	1.412

Reducing Variability Increases Profit at Constant Setpoint for Nonlinear Objective Function

Variability Reduction

- Caution - Variability reduction gives the option or opportunity for changing operating point, it does not insure operations changed to achieve benefits
- If no built-in enforcement of new changed operating point, need to discount potential benefits

Summary

- Reducing variability often has a direct operational performance economic payout
- This expected payout can be quantified and the actual payout audited
- There is a well developed methodology for analyzing and solving variability problems

Questions? Comments? Other References

1. More Material:
   http://www.emersonprocess.com/solutions/advancedautomation
2. Questions? – contact
   doug.white@emerson.com

6d GC- the to DO and the NOT to DO List for Plant Engineers

Julie Freelin, Analytical, Shell Global Solutions, 3333 Highway 6 South, Houston, TX 77082

Often, it is either underestimated and/or overestimated the capabilities of the GC as a main analytical tool for a Plant. This presentation will give an overview of how to best utilize the GC for several areas of technical support in a Plant.

6e On Electron Beam Additive Manufacturing Process for Titanium Alloys

Sharma, Renganathan Kal, Department of Chemical Engineering, Prairie View A & M University, PO Box 519, MS 2505, Room 201A, C. L. Wilson Building, Prairie View, TX 77446

Aircraft manufacturing cost and cycle times are adversely affected by titanium forgings. Additive manufacturing of titanium structural parts has the potential to reduce lead and cycle times, work in process, and the amount of titanium required to produce structural parts. Spares can be produced during small production runs and to produce prototypes where tooling or conventional manufacturing costs would be prohibitive. Advanced electron additive manufacturing, eBAM techniques and control processes to effect flight worthy titanium alloy structural aerospace components is evaluated in this study. Sensors needed for closed loop control of teh deposition parameters critical to the quality of the component are designed. A detalied case study is shown where eBAM solution is optimized and are competetive with existing titanium forging process in terms of cost, quality, lead and cycle times, WIP, etc.

A manufacturing process development plan, feasibility study, of using eBAM processes are presented. The MPD includes progress and completion performance goals and transition plan to a production partner. Scale-up, optimization and demonstration issues of representative components are discussed. How much better are eBAM parts compared with forged parts that can be produced at similar or better rates is outlined. There is a tutorial on electron beam welding at http://sbirsttrmall.com/Library/Default.aspx

6f On Nanoscale Bolometer to Study Chemical Plumes

Sharma, Renganathan Kal, Department of Chemical Engineering, Prairie View A & M University, PO Box 519, MS 2505, Room 201A, C. L. Wilson Building, Prairie View, TX 77446

Nanotechnology is rapidly emerging as a distinct discipline by itself. Nanobolometers can be designed for Infra Red and Millimeter waves by better understanding the role of electrons, molecules, photons and phonons. Imaging of chemical plumes can be performed using these nanobolometers. Current microbolometer cannot afford a multispectral/polarimetric ability and they provei for lower sensitivity and slower response times. Focal Plane Arrays, FPA, based on cooled HgCdTe photodiodes can provide an alternative. Multispectral/polarimetric ability and higher sensitivity can be expected using nanobolometers. Nanoengineering of two dissimilar materials i.e., a metal nanoparticle array in a dilectric matrix at a length scale of about 10 nm, produced a metamaterial that possess interesting mechanical, optical, electrical and thermal properties. The design and performance issues of a nanobolometer is discussed in this study.

32a So Many Good Ideas...So Little Time and Money (Webcast Not Available)

Shawna S. Berg, Process Engineering and Evaluations, Shell Global Solutions (US) Inc., Westhollow Technology Center, P. O. Box 1380, Houston, TX 77251

Though for some operating facilities it is hard to believe that there could be a day when the plant is running as it should, these days do arrive from time to time. "Good" days in the operating world are a chance for engineers to make improvements in manufacturing by cutting costs, improving performance, reducing environmental footprint, enhancing product quality, etc. More often than not, there are more great ideas than there is time and money to implement them. This session will discuss tricks of the trade for seeking out potential plant improvements, methodologies for prioritizing work economically and strategically and making the most of your budget.

32b Methods for Prioritizing Projects --- An Historical Perspective

Jonathan Worstell, Hoder, Shell Global Solutions US, 3333 Highway 6 South, Room CP-109, Houston, TX 77082 and John H. Worstell, Engineering and Construction, TST Infrastructure LLC, 61 Inverness Drive East, Suite 100, Englewood, CO 80112.

Since the mid-1950's, many methods have been devised and proposed for prioritizing technical, engineering projects. This talk presents an overview of these methods and discusses the strengths and weaknesses of each method. This talk also discusses the strengths and weaknesses of the currently preferred method.

32c Computational Economy in Process Engineering

Jed Moyano Bellen, Process Technology Division, Fluor Daniel Inc. Philippines, 3rd Floor, Asian Star Building, 2402-2404 ASEAN Drive, Filinvest Corporate City, Alabang, Muntinlupa, Philippines

Process design involves many calculations like simulation, vessel sizing, heat exchanger rating and tray rating. These calculations often involve lengthy solutions, repetitive runs and sometimes untractable convergence. To avoid this complexities, this paper explores on the concept of "Computational Economy" to put in place a practically shortened calculation time. This concept basically applies a paradigm that for certain process engineering calculations, there would be first, an excel-based program to estimate major inputs to various softwares and after which the values are tabulated to avoid using the same values over and over again. In applying this method, circuitous and repetitive calculations are avoided and the optimal values are thus obtained.

Key Words: computational economy, repetitive calculations, excel-based program

32d Using ConcepSys to Optimize Small Project Capital Allocation and Teach Front-End Design

Adnan A. Siddiqui, ConcepSys Solutions LLC, 5700 Northwest Central Drive, Ste. 301, Houston, TX 77092

ConcepSys is a new tool for integrated front-end design of projects. It integrates process flow diagram development with equipment layout and specifications to deliver front-end design and factored estimates. In this presentation, a sample project case study will be presented on how ConcepSys can improve capital allocation decisions for running plants as well as be used to teach college students and recent graduates on project front-end design. The full front-end work flow up to FEL-2 will be shared starting with developing a simulation, importing it in ConcepSys, generating preliminary design documents (equipment layout, equipment list, line list, equipment specifications) and refining the factored cost estimate. Industry practitioners in the audience will be able to rapidly evaluate and estimate project options for running plants while academia audience will be able to improve the effectiveness of senior design courses.

32e Near-Zero Flaring Study for Chemical Plant Sustainability

Qiang Xu, Kuyen Li, Jian Zhang, Chao-Wei Liu, and Xiongtao Yang. Department of Chemical Engineering, Lamar University, Beaumont, TX TX 77710

During chemical plant turnaround operations (plant startup and shutdown) and process upset management, off-spec product streams will be produced, which usually have to be sent for flaring. Flaring is a passive and end-of-pipe means to protect personnel and equipment safety in chemical process industry (CPI). However, flaring inevitably generates huge amount of CO, CO2, NOx, and some extremely hazardous volatile organic compounds (VOCs), which will cause severe air pollution problems and negative societal impacts. Meanwhile, it also results in tremendous raw material and energy loss that could generate much needed products from the industry. Thus, flare minimization has great significance on environmental, societal, and economic advancement toward the sustainability of chemical process industry.

This paper addresses planning, scheduling, and process design opportunities on near-zero flaring for chemical plant turnaround operation to advance current endeavors on flare minimization. The key idea is to reduce the startup time and pursue the "balanced" reuse of off-spec products from plant-wide systematic point of view. A general methodology framework has been developed. As a case study, the near-zero flaring thrusts for an ethylene plant startup have been studied.

Paper 32F Compact Heat Exchange Reactors for Plant Improvements

Mark Wood (1), Zhijun Jia and Larry Stryker (2). Chart Energy & Chemicals, Wolverhampton, UK (1) and LaCrosse, WI (2)

Process Intensification

Performance of chemical reactors can be improved incrementally by modifications to existing equipment, or radically by applying the concept of Process Intensification (PI) to introduce the use of new chemical reactor technologies. The PI approach designs the reactor to meet the needs of the chemical reaction, rather than fitting the chemical reaction into existing, standard equipment. By doing this, the reaction rate becomes limited by the reaction kinetics, rather than by physical actions such as heat and mass transfer. When the reaction is not physically limited, the residence time is then minimized and the chemical reactor becomes much smaller, minimizing inventory and ensuring consistent, high product quality and yield. Improving the reactor will benefit the entire plant, possibly reducing the need for downstream separation or purification stages.

When correctly designed, a PI process can achieve any or all of the following benefits:

- Improved product quality
- Reduced residence time
- Reduced reactant inventory
- Reduced energy consumption
- Reduced capital cost for plant modifications
- Inherently safe operation

PI is of particular benefit for faster chemical reactions that are limited by heat transfer or mass transfer and when operating in a continuous manner. It is recognized that not every chemical reaction will benefit from PI due to the nature of the kinetics or physical form of the reactants and products, but despite this, a number of improvements could still be made by applying the principles of PI. Using PI reactor technologies can allow reactions to be operated under conditions that simply would not be possible within existing equipment.

Chemical reactions are traditionally carried out in stirred tanks, columns and tubular reactors. Process scale-up is achieved by increasing the dimensions of the equipment to give the required production rate. When scaling up, the surface area per unit volume decreases, meaning the heat addition or removal capability is decreased when increasing the scale. This gives rise to differing performance between the prototype, pilot and full scale operation and is particularly problematic for exothermic reactions where increased temperature can promote unwanted byproduct reactions and even potentially cause runaway conditions. Techniques used to work around limited heat transfer capability include using inert diluents, reducing feed rate for stirred vessels, or having inter-stage cooling for continuous processes. All of these options introduce additional costs to the process.

With continuous operation of a fast exothermic reaction, there will be a rapid increase in temperature when the reactants are mixed. This temperature peak can be very undesirable if it causes product degradation or safety issues. PI reactors can be designed to precisely add one component to another in a specific location, or in multiple locations, ensuring that the temperature peak can be minimized when coupled with their high heat transfer capability.

Mixing is also affected as equipment increases in size. Mixing rate is a function of the energy dissipation within the reactor fluid, typically measured in watts of mixing energy per mass of fluid (w/kg). As physical size increases, it becomes more difficult to achieve the same energy dissipation. This can affect the product quality for fast reactions if the increased mixing timescale gives the opportunity for slower byproduct reactions to occur. By applying PI, mixing capability is maintained between different operating scales, primarily by scaling up to a larger number of channels of the same dimension, rather than having a similar number of larger channels.

Compact Heat Exchange Reactors

The principle of operating chemical reactions inside heat exchangers is one means of achieving PI. As heat exchangers are designed to have high heat transfer capability, they are suited to the task of providing and removing heat of reaction. Compact heat exchangers have short heat transfer paths and high heat transfer surface areas per unit volume, making them suitable for effectively removing or adding heat to chemical reactions. It is important to note that "Compact" refers to the spacing of the channels and distance between heat transfer surfaces, rather than the physical size of the unit, as these exchangers and reactors can be very large.

Compact heat exchange reactors are a form of PI reactor that act as an Enabling Technology, allowing operators to run under conditions that were previously unobtainable. Compact heat exchange reactors can have a very high surface area per unit volume when compared to conventional reactors. Figure 1 illustrates the surface area per unit volume of different reactors, measured as m^2 of surface area per m^3 of fluid volume (m^2/m^3). Values of 10,000 m^2/m^3 relate to channels with dimensions of around 0.1mm. Typical operating dimensions are in the mm scale, giving 2000-4000m^2/m^3.

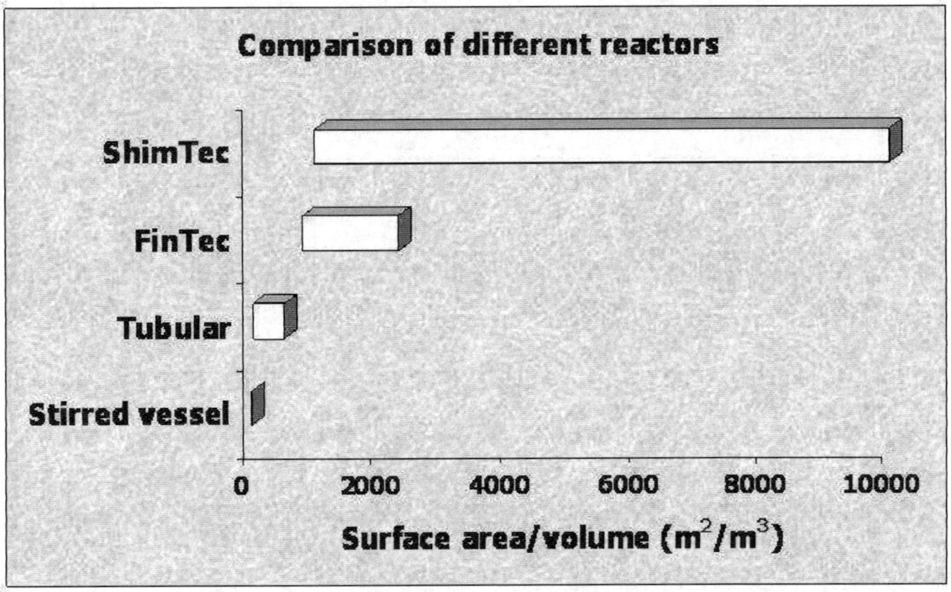

Figure 1. Surface area per unit volume of different reactor options

Chart manufactures two ranges of compact heat exchange reactors, FinTec® and ShimTec®. FinTec® is suitable for medium to large scale operation, whilst ShimTec® is suitable for small to medium scale operation. A particular version of ShimTec® has previously been referred to in literature as "Marbond" and "HEX". Both ranges can be classed under the umbrella of "structured reactors" as the internals are precise geometric structures that are readily replicated throughout the reactor. Figure 2 illustrates the range of physical dimensions achievable by Chart's heat exchangers and heat exchange reactors.

ShimTec reactor

Plate-fin heat exchanger

Figure 2. Different scales of compact heat exchangers manufactured by Chart

Chart has been manufacturing compact brazed aluminum plate-fin heat exchangers for over 50 years, with these being used as chemical reactors as early as 1979. Secondary heat transfer surfaces are manufactured from material formed into fins and brazed into an overall structure. FinTec® is the name Chart applies to plate-fin reactors. In brazed aluminum, these reactors can have volumes of several cubic metres, making them suitable for bulk chemical processing. Aluminum is an appropriate material of construction for a number of chemical reactions, with operation at temperatures up to 260°C and pressures up to 30barg. Heat transfer capability, measured as UA, is about 20 times greater than a shell and tube heat exchanger, allowing very good temperature control in reactions. Stainless steel plate-fin reactors can also be manufactured.

ShimTec® is a more recent development, though still dates back more than 15 years. Construction is from thin metal sheets with features cut into them, stacked to provide channels. The robust structure lends itself to being diffusion bonded, resulting in a reactor capable of withstanding extremely high operating temperatures and pressures. Reactors have been operated at well over 800°C and proven to withstand design pressures of 180 bar at lower temperatures. ShimTec® reactors are available in a range of metals including stainless steel and nickel alloys.

Both reactor types are designed around the same principle with a number of alternating layers, so that reaction streams are sandwiched between heat transfer streams. Intermediate layers and channels can be incorporated to allow the controlled addition of additional chemicals into the bulk streams. This addition can be divided up along the length of the reaction path, ensuring that temperature profiles or product compositions are closely controlled (Figure 3).

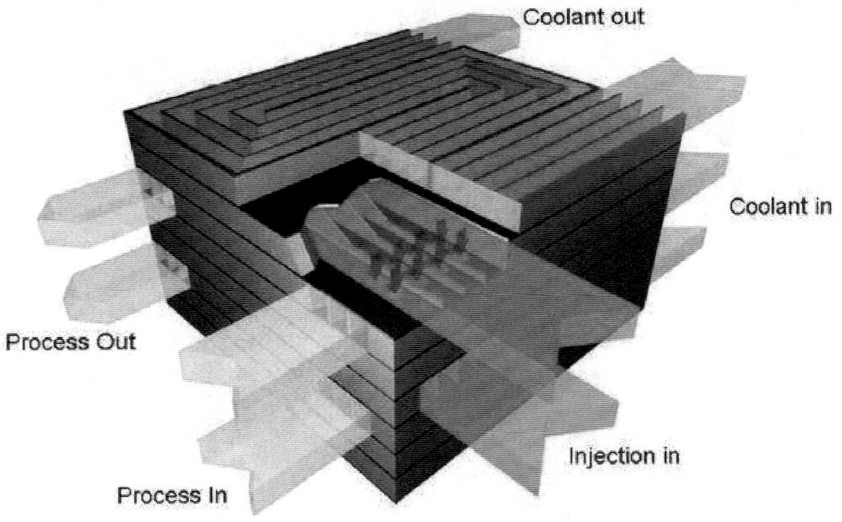

Figure 3. Layer arrangement of ShimTec and FinTec reactors

Both FinTec® and ShimTec® are readily scaled up as the channel dimensions can be kept constant, regardless of the physical size of the finished unit. This avoids the scaling-up difficulties seen in conventional equipment.

Catalyst Incorporation

Catalysts are widely used within the chemical process industry. Consequently it is important that catalysts can be incorporated within the generally smaller passageways of compact heat exchange reactors. Chart has used a variety of techniques for the inclusion of catalyst into the reactor structure:

- Coated on internal surfaces
- Packed
- Pre-applied to a removable insert
- Injected along the reaction path

Open channels with dimensions of several millimetres upwards can be produced to allow packing and insertion of catalyst, which can be removed and replaced when required.

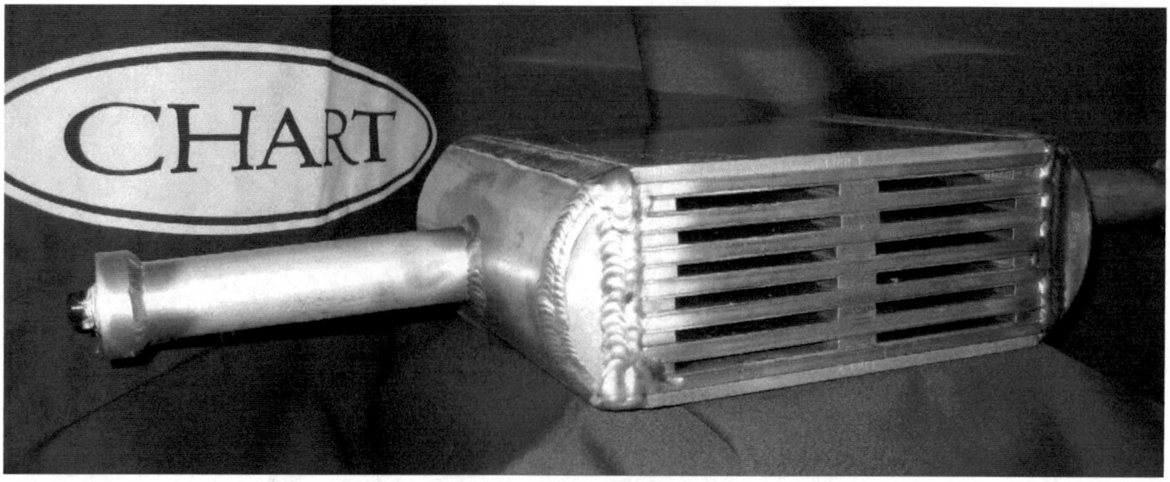

Figure 4. Open channels for packing with catalyst

Coating catalyst allows the reaction to take place immediately adjacent to the metal structure, providing very short heat paths.

How to consider when PI is applicable

Processes may be running successfully in their existing reactors, but there are a number of questions that can be asked to consider whether the application of PI can benefit the reaction. Questions to ask when considering whether a chemical process can benefit from PI are:

- Is product quality (yield, conversion) poorer at the full scale than at laboratory or pilot scale?
- Is the residence time longer at full scale than at smaller scales?
- Are inert fluids added to the process stream to provide a thermal mass, and/or to slow down the reaction kinetics?
- Is inter-stream cooling required to supplement heat transfer from the equipment itself?
- Does catalyst degrade in specific locations, suggesting the presence of hot spots?
- Do operators feel like they are putting the brakes on the reaction to slow it down?
- Are there any alternative process routes (e.g. reactants) that people would like to try, but do not dare to in the existing reactor equipment?

If the answer to one or more of the above questions is yes, then it is possible that PI will benefit the process. Techniques exist for assessing the process further to determine how to apply PI to the process[2]. In essence, the reaction kinetics are examined and studies carried out to determine the ideal conditions for the reaction to operate under, looking at temperature and pressure of operation, heat and mass transfer requirements and how chemicals are to be added together.

Using these PI principles, an example case study has been carried out in a Chart ShimTec reactor[3]. The process operates in a semi-batch stirred vessel and is an exothermic reaction with a liquid-phase catalyst, producing significant quantities of byproducts. Feed addition rate has to be restricted to control the reaction rate. A Chart compact heat exchange reactor with a high heat transfer capability allowed the operating temperature to be increased to enhance reaction kinetics and reduce the required residence time. Fluids were pre-mixed with a rotor-stator mixer to generate a two-phase dispersion, before entering the channels which maintained the dispersion. Reaction time was reduced from 1 hour to 6 seconds with a 99% reduction in byproduct formation.

Conclusions

Process Intensification of chemical reactions through the use of compact heat exchange reactors has been proven to provide great benefits. Chemical processes have key characteristics that can be examined to determine whether PI can be applied to the process. Compact heat exchange reactors are suitable for intensification of generally fast chemical reactions and particularly those that are limited by heat transfer.

References

1. Tsouris and Porcelli. "Process Intensification – Has its time finally come?" CEP October 2003, 50-55.

2. Reay, Ramshaw and Harvey (2008). Chapter 12 of "Process Intensification" Elsevier Ltd, Oxford, ISBN 978-0-7506-8941-0

3. Green, Johnson, Westall, Bunegar and Symonds. "Combined Chemical Reactor/Heat Exchangers: Validation and Application in Industrial Processes." 4th International Conference on Process Intensification for the Chemical Industry, Brugge, 10-12 September 2001.

Surfactant EOR Process Development Fundamentals

Joseph B. Powell, Ph.D.
Chief Scientist – Chemical Engineering
Shell Global Solutions U.S. Inc.
Westhollow Technology Center, Houston, TX

Overview

Primary (pressure drive) and secondary (waterflood) oil recovery produces less than 50% of the reservoir's original oil in place (OOIP). The remaining oil is trapped by capillary forces which arise from the naturally high interfacial tension between oil and water phases. The interfacial tension (γ) is a measure of the thermodynamic penalty that must be overcome to alter an oil droplet from the minimum area of a sphere, in order to squeeze through a pore throat and travel towards the production well (Figure 1). For oil-water systems with interfacial tensions on the order of 10 dynes/cm, the deformation price is high, and water flows around the trapped oil droplets. The latter thus comprise a residual oil saturation (S_{OR}) of around 25 – 30% of the pore volume at the end of normal production.

Surfactants may be added to the injected water to reduce interfacial tensions and recover trapped oil (Figure 2). *Tertiary recovery* in this manner is attractive, because the location of the oil is well known from previous production, and existing wells may be used, if not abandoned. Since *enhanced oil recovery* (EOR) has rarely been economic relative to new oil exploration and production, the total quantity of remaining oil more than exceeds cumulative global total production, to date.

Design of a surfactant-flood EOR process

In order to improve recovery, viscous forces must overcome capillary trapping forces, as reflected in the dimensionless capillary number N_C (Figure 3). While water can be thickened by polymer to improve viscosity and areal sweep, viscous forces are limited by injection capability and provide only limited value in increasing recovery beyond that which can be effected by water flooding. Modification of interfacial tension by use of surfactants to reduce the capillary trapping force, is the critical control parameter for achieving high tertiary recovery. Interfacial tensions must be reduced from 10 to at least ca. 0.01 dynes/cm to mobilize residual oil (Reed 1977). Reduction to 1×10^{-3} dynes/cm is considered "ultralow" and a target to insure complete recovery.

Design of surfactants to produce low or ultralow interfacial tension is a function of oil composition, such that each crude oil requires separate design of an optimal surfactant or surfactant blend (Puerto 1982). The conditions giving rise to low interfacial tensions require delicate balancing of the hydrophobe (lipophilic) and hydrophile character of the surfactant. Temperature and water composition (salinity and divalent ion content) are also important. Optimal balancing may be a function of surfactant concentration, and vary throughout the surfactant flood.

While there is some limited ability to predict optimal surfactant structures *a priori*, empirical screening for low interfacial tensions using a structured salinity scan approach is used for process design (Puerto 1982). Salinity-scan phase behavior tests are conducted in graduated tubes (10 –ml) where target crude oil is mixed with a test surfactant blend, with variation in brine salinity centered at target reservoir salinity. The tubes are rotated (Figure 5a) in a thermostatted oven for at least overnight to achieve equilibrium, then allowed to settle. Volumetric measurements are then taken to assess the amount of oil solubilized in the aqueous surfactant phase at low salinity (V_o/V_S), and the amount of water solublized in the surfactant-laden oil phase at high salinity (V_w/V_S). Salinity will drive surfactant partitioning from lower phase to the upper phase, through a middle phase of optimal low interfacial tension (Figure 5b). The region of low interfacial tension is typically quite sharp, requiring delicate balancing of surfactant design and performance (Figure 6). Low interfacial tensions (0.01 dynes/cm or less) required for enhanced oil recovery generally correspond to a solubilization parameter V_o/V_S or V_w/V_S of 10 or higher (Reed 1977), with underlying theory derived (Huh 1982).

Phase measurements can be problematic with opaque crude oils. Brines used for salinity scans should include multivalent cations as found in the target reservoir. Once an optimal surfactant blend has been identified, including possible use of co-solvent to obtain miscibility and polymer for mobility control in the "surfactant slug", low interfacial tensions are confirmed directly. A spinning-drop interfacial tensiometer is ideal for low-tension systems. Viscosities of crude oil, surfactant blends, and polymer drive banks are measured. Polymer is added to the surfactant blend composition to obtain a mobility ratio M less than one, such that the mobility of the displaced phase (e.g. "1"= oil) is greater than that of the displacing phase ("2" = surfactant), and similarly for polymer slug chasing surfactant. Here, "mobility" M is defined as the phase relative permeability divided by the viscosity.

$$M = \frac{k_{r2} / \mu_2}{k_{r1} / \mu_1}$$

Mobility control in this manner reduces potential for bypassing due to viscous fingering of one phase through another, to give optimal contacting. Optimal process design thus requires knowledge of oil properties and reservoir characteristics.

After surfactant and polymer selection, the concept is demonstrated in a laboratory coreflood. Generic sandstone (Berea) or carbonate may be used by pre-evacuating the core, filling with representative brine, driving via oil flood to irreducible (connate) water saturation (15 – 25% pore volume PV), then waterflooding to obtain residual oil saturation (S_{OR}). At this point, the oil saturation in the core will generally range from 25 – 35%, and no more oil can be produced by further waterflooding. Application of a successful surfactant slug will drive residual saturations relative to chemical flooding (S_{OC}) quite low: Greater than 90% recovery is generally possible for rock zones swept by surfactant in laboratory cores. The coreflood may be repeated with preserved native core from the target reservoir, as final confirmation.

Process design entails minimizing surfactant slug size and cost. Designer surfactants to achieve ultralow interfacial tensions are more than an order of magnitude more expensive than the value of the oil produced. Surfactant may be lost via adsorption on reservoir rock, particularly clays, and by phase trapping or formation of oil-soluble phases, which for some microemulsion structures may be high in viscosity. Alkaline agents may be used to generate in-situ surfactants for sacrificial adsorption. Polymers must be salinity and divalent ion tolerant, and be thermally stable under intended flood conditions. Polyacrylamide polymers are subject to shear degradation, such that injectivity tests must be conducted. Biopolymers (xanthan gum) are subject to subsurface biodegradation.

Reservoir heterogeneity must be characterized and modeled. The intent of surfactant flood design is to inject the minimum slug of surfactant required to just sweep the reservoir prior to exhaustion by adsorption. Variation in surfactant concentration and slug size impacts timing of oil recovery, and mobility ratio of transition zones where polymer and surfactant may propagate at different velocities due to differential adsorption or phase trapping. Adsorption under reservoir conditions for surfactant and polymer must therefore be assessed. Heterogeneity can cause channeling of surfactant through thief zones directly to the production well, leading to low recovery. Process development for surfactant flooding is therefore an iterative process relying on reservoir characterization and modeling, field injection tests, and laboratory design of efficient surfactant systems that will mobilize oil for those portions of the reservoir contacted.

Figure 7 details the essential features of a commercial surfactant flood. A surfactant slug of optimal concentration and viscosity is injected at a pore volume less than half of the target displacement. This is chased by a lower cost and lower mobility polymer bank, to drive the surfactant through the reservoir at optimal sweep efficiency. Water may be used to chase the polymer bank, though viscous fingering will be problematic due to lower viscosity. Salinity gradients may be employed (Hirasaki 1980) before or after the flood to reduce salinity and or divalent ion content, if improved phase behavior and reduced surfactant retention is indicated.

References

Hirasaki, G. J., and H. R. van Domselaar, R. C. Nelson. 1980. Evaluation of the Salinity Gradient concept in Surfactant Flooding. *SPE 8825.*

Huh, C. 1982. Equilibrium of a Microemulsion that Coexists with Oil or Brine. *SPE 10728.*

Puerto, M. C., and R. L. Reed. 1982. A Three-Parameter Representation of Surfactant-Oil-Brine Interaction. *SPE 10678.*

Reed, R. L, and R. N. Healy. 1977. Some Physicochemical Aspects of Microemulsion Flooding: A Review. In *Improved Oil Recovery by Surfactant and Polymer Flooding*, edited by D. O. Shah, and R. S. Schechter. NY: Academic Press.

Figure 1: Capillary Trapping in Pore (Young-Laplace Equation)

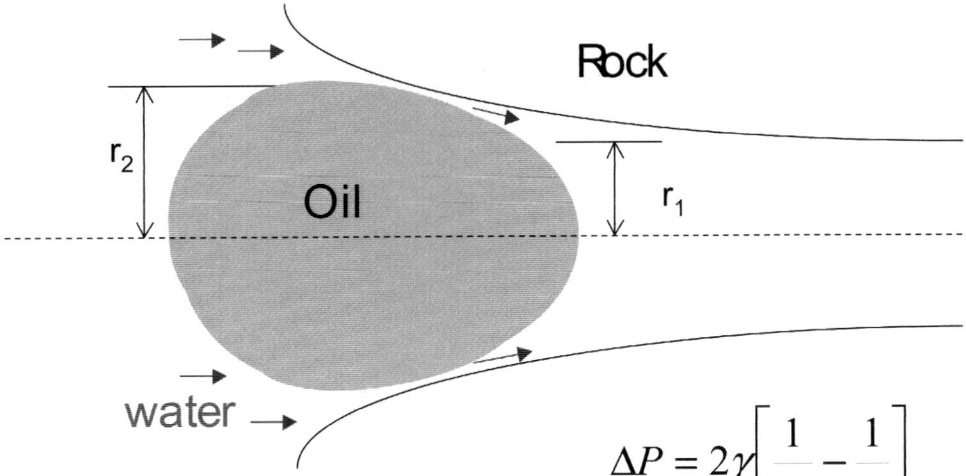

$$\Delta P = 2\gamma\left[\frac{1}{r_1} - \frac{1}{r_2}\right]$$

Figure 2: Surfactant adsorption at oil-water interface to reduce interfacial tension

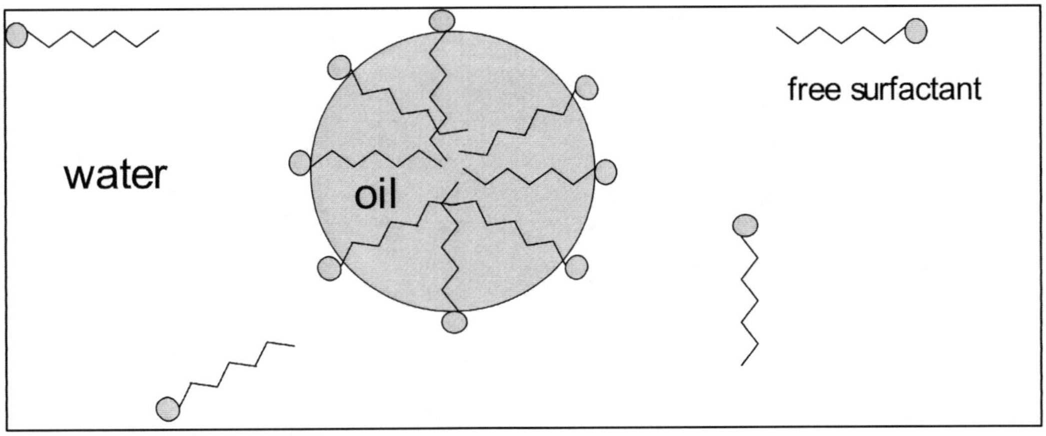

Figure 3: Remaining oil saturation vs. capillary number for surfactant flood

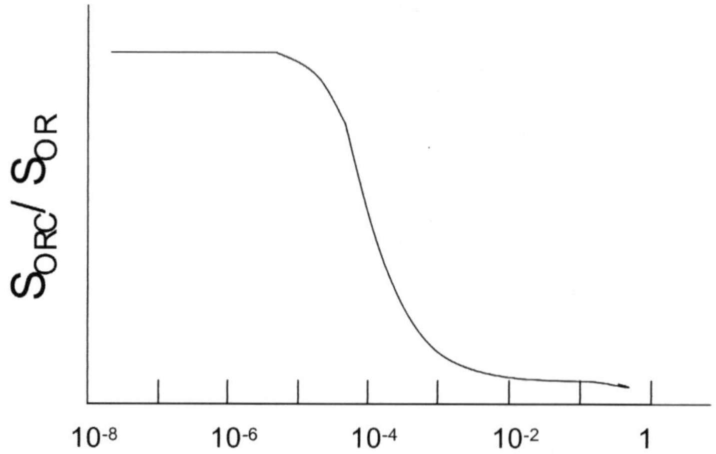

$$N_C = \frac{u \bullet \mu}{\gamma}$$

Figure 4: Phase behavior vs. Salinity (Windsor)

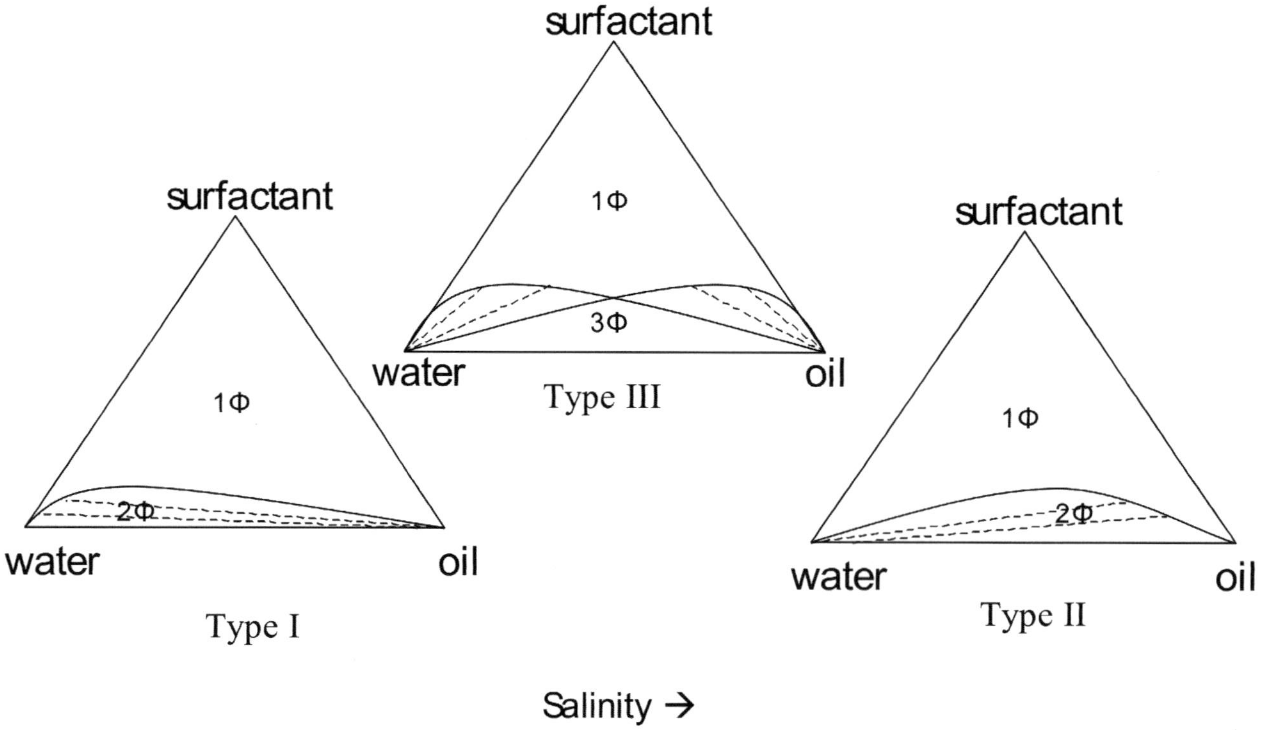

Figure 5: Phase behavior experiments

(a) rotating rack for equilibration

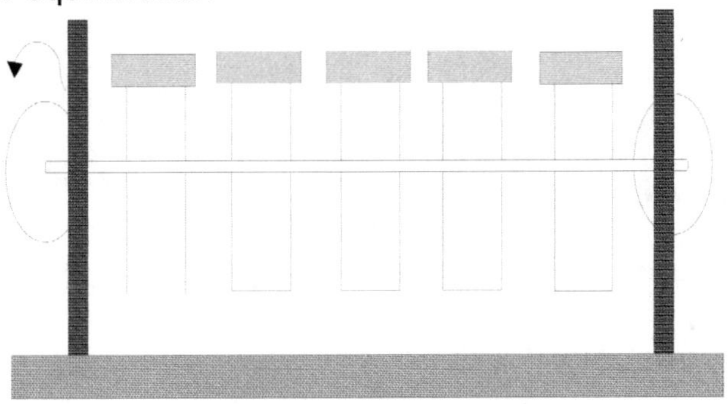

(b) settling & phase measurement

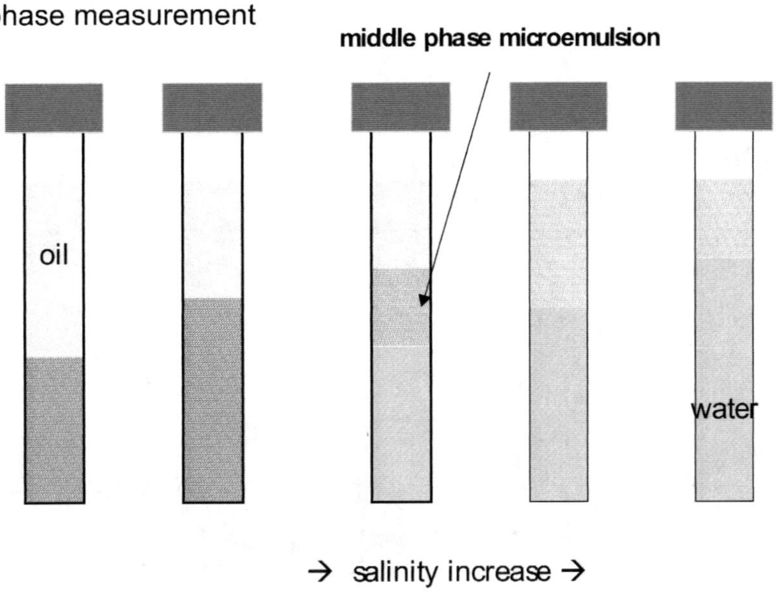

Figure 6: Ultra-low interfacial tensions and solubilization parameter Illustration from (Reed 1977).

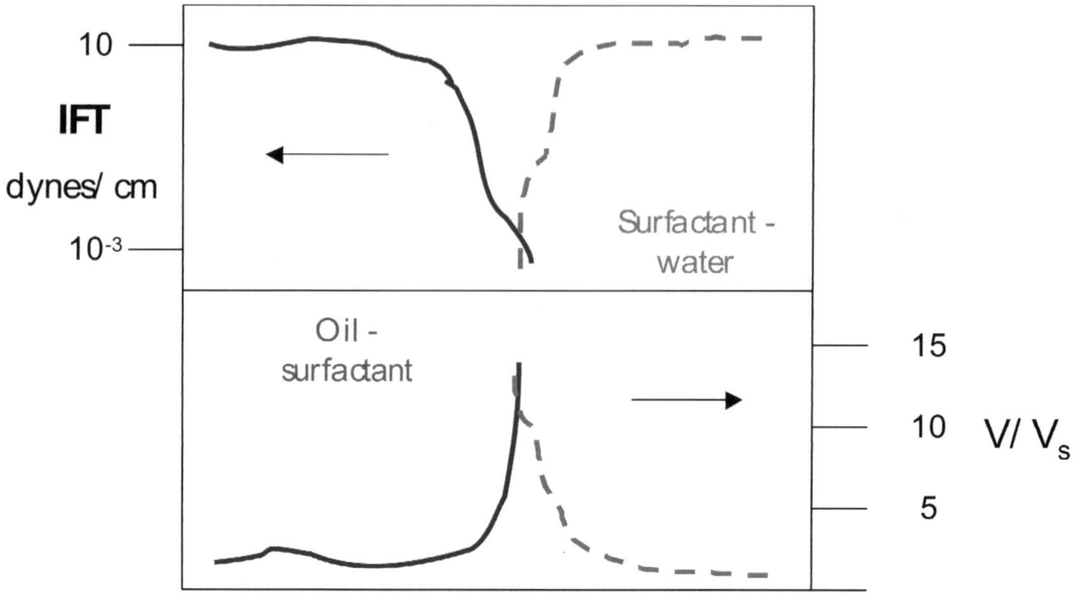

Figure 7: Surfactant Flood Design

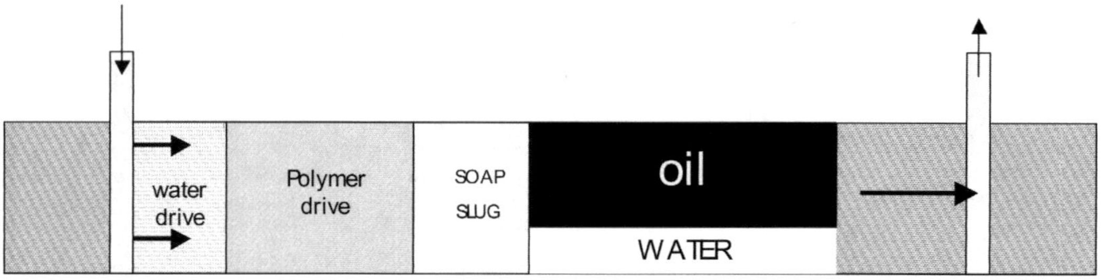

- Oil is banked ahead of surfactant slug used to reduce interfacial tension.
- Polymer increases viscosity / reduces "mobility" to prevent channeling

53b Producing Chemicals for the Enhanced Oil Recovery Market

Jonathan Worstell, Hoder, Shell Global Solutions US, 3333 Highway 6 South, Room CP-109, Houston, TX 77082 and **Mike J. Doll**, HODer, Shell Global Solutions, 3333 Highway 6 South, Houston, TX 77082.

The chemicals used for Enhanced Oil Recovery are speciality chemicals produced in commodity quantities. This talk discusses this dichotomy and how to produce such products via contract manufacturing. The "ins and outs" of contract manufacturing will also be presented in this talk.

53c The Economics of Chemical Enhanced Oil Recovery (Webcast Not Available)

Shawna S. Berg, Process Engineering and Evaluations, Shell Global Solutions (US) Inc., Westhollow Technology Center, P. O. Box 1380, Houston, TX 77251

Though there are many strategic reasons to produce or utilize enhanced oil recovery chemicals, businesses must decide when pursuing this option is the most profitable. Every oil field is different. The price of the barrel changes daily. How can businesses make the best long-term choices for their bottom line? This session will discuss the methodology and mindset of assessing the economics of chemical enhanced oil recovery from production of EOR chemicals to utilization in the field.

Dynamic Effects for Two-Phase Flow in Porous Media "Temperature Effects"

Dr. Navraj Hanspal, University of Manchester, UK
Dr. Diganta Bhushan Das, Loughborough University, UK

Abstract

Traditional continuum scale models for multiphase flow in porous media are based on "*Capillary Pressure, Saturation & Relative Permeability [P_c-S-K_r]*" relationships which do not necessarily incorporate the dynamic capillary pressure effects on the flow behavior. To address this issue, relationships with the inclusion of a dynamic coefficient(s) have been proposed that are shown to have important practical effects, which increases as the scale of the domain increases. As such, simulators for multiphase flow in porous media must include additional terms(s) associated with dynamic capillary pressure. For these reasons investigations targeting at measurements of dynamic coefficients and its dependency on various physical parameters are of great interest. Such investigations aid in developing upscaling relationships that can be cost-effectively used for characterizing dynamic multiphase flow behaviour. In this work we examine the dependence of the dynamic coefficient (τ) on temperature (T) by carrying out quasi-static and dynamic flow simulations for an immiscible perchloroethylene (PCE) - water system exhibiting a drainage process. Simulations are carried out for a range of temperatures between 20^0C-80^0C on 3-D cylindrical domains which correspond to laboratory scale domain set-ups used previously by the authors. The numerical data are fitted to the Dynamic Capillary Pressure(P_c)-Saturation(S) curves for quantifying the (τ) values and later for examining the lumped dynamic effects arising from the physical properties and temperature variations in two different soil domains i.e. coarse and fine sands. Results are interpreted by examining the correlations between dynamic coefficient(s) and temperature, time period(s) required for attaining residual saturation and the dynamic aqueous/non-aqueous phase saturation and capillary pressure plots. It is envisioned that our simulations maintain a continuity of our previous work, reduce the inconsistencies associated with the dependency of dynamic coefficient(s) on temperature which will subsequently enable us to carry out computationally economical and more accurate modelling studies at various scales of observations.

1.0 Introduction

"*Capillary Pressure, Saturation & Relative Permeability [P^c-S-K_r]*" used in conjunction with the extended form of Darcy's Law for multiphase flows (Muskat *et al* 1932; Scheidegger 1974, Whitaker 1986a and Hanung 1997) play an important role in quantitative modeling of multiphase flows in porous media. These P^c-S-K_r relationships are highly non-linear in nature and depend on flow hydrodynamics (dynamic/static) conditions, capillary or viscous forces, contact angles, grain size distribution, micro-scale heterogeneities, surface tension, boundary conditions, fluid properties, and length scales of observation. The conventional empirical macroscopic relations of the capillary pressure are functions of wetting phase saturation (P^c-S relationships) and differences between the average pressures for oil/non-wetting and the water/wetting phase (Bear & Verrujit 1987; Helmig 1997). These relationships can be mathematically expressed as,

$$P_{nw} - P_{w} = P^{c,equ}(S) = f(S) \qquad (1)$$

where P_{nw} is the average pressure for oil/non-wetting phase and P_w is the average pressure for water or wetting phase, S refers to the water or wetting phase saturation. Equation (1) exhibits a linear-relationship between the saturation and capillary pressure. This relationship is obtained under steady-conditions in fluid equilibrium when saturation remains unchanged with time, considering all the aspects that affect the flow behavior in porous media such as; material properties (e.g., porosity, intrinsic permeability, pore size distribution, etc), fluid properties (densities, viscosities, interfacial tensions, etc.), contact angles, wettability and the media heterogeneity. Steady state *[P^c-S-K_r]* relationships have been studied by several authors (Beliaev & Hassanizadeh 2001; Beliaev & Schotting 2001; Murdoch & Hassanizadeh (2002); Hassanizadeh *et al.* 2002; Hanyga & Seredynska 2005) to discuss the capillary hysteresis curves in dynamic two-phase porous flow situations with thermodynamic restrictions and verifications to physical modelling. They conclude that the conventional steady state *[P^c-S-K_r]* relationships are not sufficient to characterize the behavior of two-phase flows in porous media, as they fail to predict the dynamic flow, dynamic capillary pressure effects, dynamic temperature effects and interaction of fluid/fluid interfaces arising from various capillary, viscous forces and changes of saturation over prolonged time periods.

In response, Hassanizadeh & Gray (1993) were the first ones to propose the generalized P^c-S relationships with the inclusion of dynamic capillary pressures. They concluded that dynamic effects do not occur in steady-state flows and as a result, the definition of macroscopic capillary pressure should be based on thermodynamic considerations. The main hypothesis is the dependence of interfaces variables (namely; mass density, temperature, saturation, porosity, etc) and co-existance of phases on Helmholtz free energy. As a result, the generalized P^c-S relationships with the inclusion of dynamic capillary pressure can be written as,

$$\left(P^{c,dyn} - P^{c,equ}\right)\Big|_S = -\tau\left(\partial S / \partial t\right)\Big|_S \qquad (2)$$

where τ is the dynamic coefficient, $\partial S / \partial t$ is the time derivative of saturation, $P^{c,dyn}$ is the dynamic capillary pressure, $P^{c,equ}$ is the steady-state capillary pressure and the measurements for all variables is taken at the same saturation value. The dynamic coefficient can be determined from the slope of Equation 2, which exhibits a linear relationship in form of a straight-line.

As reported in the literature, Equation (2) has been subsequently used by many workers i.e. Theodoropoulou *et al.* (2005), Gielen *et al.* (2005), Dahle *et al.* (2005), Manthey *et al.* (2005), Das & Hassanizadeh (2005); O'Carroll *et al.* (2005); Oung *et al.* (2005); Berentsen *et al.* (2006); Bottero *et al.* (2006); Mirzaei & Das (2007); Das *et al.* (2007) for quantifying the range of the dynamic coefficients (τ) and examining its dependency on the size (10^{-3}-1m), geometry (2D/3D domains), boundary conditions (entry pressure etc.), physical properties of the porous media (permeability, porosity, etc.), fluid parameters (densities, viscosity ratios etc.), types of heterogeneities (uniform, random or micro-scale), flow characteristics (Newtonian, non-Newtonian etc.) and gravitational effects. Based on the preliminary works of Stauffer (1978), more recently Tzakiroglou *et al.* (2003), Dahle *et al.* (2005), Manthey *et al.* (2005), have proposed upscaling relationships of the form $\tau = f(\varphi, \mu_{nw}/\mu_w, \rho, K, g, \lambda, Ca, P^d)$ where τ is the dynamic coefficient, φ is the porosity, μ_{nw}/μ_w is the ratio of the viscosities of the non-wetting to wetting phase, ρ is the density, K is the permeability tensor, g is the acceleration due to gravity, Ca is the dimensionless capillary number, λ *and* P^d are Brooks-Corey (1968) parameters. Other state-of-the-art approaches proposed in literature for quantifying the uncertainity of flows in porous media include the pore-scale modelling techniques developed by Kueper & Frind (1989), Blunt *et al.* (1991), Schroth et al. (1998), Nordhaug *et al.* (2003) and Panfilov & Panfilova (2005).

In addition, the effects of temperatures on the two phase flow behavior in porous media have been only qualitatively investigated in the studies by Sinnokrot *et al.* (1971), Nutt (1981), Hassanizadeh & Gray (1990), Davis (1994), Imhoff (1995), Grant & Salehzadeh (1996), Harr *et al.* (1984), Gatmiri & Dillage (1997), She & Sleep (1998), Narsimhan & Lage (2001), Muralidhar & Shoerey (2003), Grant (2003), Hanyga & Jianfei (2003), Schembre *et al.* (2006). Although there have been phenomenal developments over the past ten years in the theory based on the inclusion of damping coefficients in P^c-S-K_r relationships, no studies till date have been reported in literature which characterize the effects of temperature on the dynamic coefficient (τ). These effects can be readily incorporated through empirical relations for fluid and material properties (interfacial tension, residual saturation, viscosity, density, relative permeabilities, etc.) as a function of temperature and can help to examine the dependencies of dynamic coefficients on temperature, which is the focus of the work presented in this article.

2.0 Mathematical Model

2.1 Hydrodynamic equations of two-phase flow in porous media

The hydrodynamics of a two phase flow in porous media can be modeled using an extended version of Darcy's equations (Muskat & Meres, 1936; Demond & Roberts, 1987) as described below,

$$\overline{q_\gamma} = -\frac{K_{r\gamma}}{\mu_\gamma}\left(\nabla P_\gamma + \rho_\gamma g \nabla z\right) \tag{3}$$

where $\overline{q_\gamma}$ is the Darcy's flux, $K_{r\gamma}$ is the relative permeability, μ_γ is the viscosity of related phase, ∇P_γ is the driving pressure gradient and ∇z is the upward unit vector. Equation 3 is applied for symmetric relative permeability tensor (isotropic) where there is no momentum transfer across the phases. It is assumed here that the flow is Newtonian and incompressible.

2.2 Continuity equation for two-phase flow in porous media

The mass balance equation for a two-phase flow in porous media can be described by the continuity equation (4) as listed below,

$$\frac{\partial}{\partial t}\left(\phi \rho_\gamma S_\gamma\right) + \nabla\left(\rho_\gamma q_\gamma\right) = 0 \quad for\ \gamma \equiv w, nw \tag{4}$$

where ϕ is the porosity, ρ_γ fluid density, S_γ is the saturation for corresponding phase, q_γ is the fluid flux, $\gamma \equiv w, nw$ refer to the wetting and non-wetting phases respectively.

2.3 Constitutive equations

Material properties such as porosity, tortuosity, compressibility or Brooks & Corey parameters are considered to be independent of temperature variations whilst the change in material particle density can be neglected (Marshall & Holmes 1973; Farouki 1986; Mandal & Divshikar 1995; Kozak & Ahuja 2005). Fluid densities, fluid viscosities, water residual saturation, and interfacial tension are assumed to be temperature variant. As such the additional constitutive relations that model the temperature effects on these properties, are described by the following equations:

2.31 Phase Saturation

The relation between the wetting phase saturation and the non-wetting phase saturation is given by equation (5),

$$S_w + S_{nw} = 1 \tag{5}$$

where, S_w is the average saturation of the wetting phase, S_{nw} is the average saturation of the non-wetting phase. The Brooks-Corey functions (6-8) are used to determine the capillary pressure-saturation relationships (Brooks & Corey, 1964).

$$S_{ew} = \left(\frac{P^c}{P^d}\right)^{-\lambda} \quad for\ P^c \geq P^d \tag{6}$$

$$S_{ew} = 1 \quad for\ P^c \leq P^d \tag{7}$$

$$S_{ew} = \left(\frac{S_w - S_{rw}}{1 - S_{rw}}\right) for\ 0 \leq S_{ew} \leq 1 \tag{8}$$

where, S_{ew} is the effective saturation of the wetting phase, S_{rw} is the irreducible saturation of the wetting phase, P^d is the displacement or entry pressure, (Nm^{-2}), λ is the Brooks and Corey pore size distribution index.

2.32 *Relative permeability*

The Burdine porosity distribution along with the Brooks-Corey formulations is used to obtain the relative permeability (9-10) of non-wetting and wetting phases (Brooks-Corey, 1964).

$$K_{rw} = S_{ew}^{(2+\lambda)/\lambda} \tag{9}$$

$$K_{rnw} = \left(1 - S_{ew}\right)^2 \left(1 - S_{ew}^{(2+\lambda)/\lambda}\right) \tag{10}$$

where, K_{rw} is the relative permeability for the wetting phase, K_{rnw} is the relative permeability for the non-wetting phase.

2.33 *Porosity*

Diffusive porosity which represents the absoluteness of the connected pore spaces is determined by equation (11),

$$n_D = 1 - (1 - \overline{n_D})\exp(-\alpha_P(P - \overline{P})) \tag{11}$$

where n_D is diffusive porosity, $\overline{n_D}$ is reference diffusive porosity, P is pressure, $\overline{P}$ is reference pressure. All connected and unconnected pore spaces are modelled as a function of total porosity using equation (12),

$$n_T = 1 - (1 - \overline{n_T})\exp(-\alpha_P(P - \overline{P})) \tag{12}$$

where n_T is total porosity. In both these equations the porous media coefficient of pressure compressibility is given by equation (13),

$$\alpha_P = \frac{S_s}{\overline{\rho_l} g} - \overline{n_D}\beta_P \qquad (13)$$

where S_s is the coefficient of specific storage, $\overline{\rho_l}$ is reference liquid density, β_P is the water coefficient for pressure compressibility.

2.34 Tortuosity

Tortuosity is determined using the Millington and Quirk (1959) function,

$$\tau_\gamma = \left(n_D\right)^{1/3}\left(s_\gamma\right)^{7/3} \quad \textit{for } \gamma = w, nw \qquad (14)$$

where τ_γ is the tortuosity of phase γ.

2.35 Residual saturation

Change in water residual saturation is described by equation (15) (She & Sleep 1998). In this work, we have used this relation in conjunction with Brooks-Corey (1964) relations.

$$S_{wr} = a_r + b_r T \qquad (15)$$

where, a_r and b_r are constants for water residual saturation function, T is temperature

2.35 Liquid-gas interfacial tension

The interfacial tension for a liquid-gas system can be mathematically modeled by equation (16) Grant (2003),

$$\gamma^{lg} = a_0^{lg} + a_1^{lg} T\ln(T) + c^{lg}\, T \qquad (16)$$

where γ^{lg} is the water-air interfacial tension, a_0^{lg} , a_1^{lg} and c^{lg} are constants for liquid and gas interfacial tension.

2.36 Liquid-liquid interfacial tension

The interfacial tension between two immiscible liquid phases can be modeled using equation (17) (She & Sleep 1998; Grant & Bachman),

$$\gamma^{lo} = a + bT \qquad (17)$$

where γ^{lo} is the water-oil interfacial tension, a and b are constants for water-oil interfacial tension

2.37 Water viscosity

The water viscosity as a function of temperature can be modeled using the empirical expression (18) described by (Yaws et al., 1976; White and Oostrom, 2003),

$$\mu_l = \exp\left(-24.71 + 4209/T + 0.04527T - 3.376\text{x}10^{-5}T^2\right)\text{s} \qquad (18)$$

where, μ_l is water viscosity, T is the temperature

2.38 Water density

Water density as a function of temperature is modelled using steam tables (Meyer *et al.* 1967;

White & Oostrom 2003) by equation (19),

$$\rho_1 = {v_c^{w}}^{-1}\{A_{11}a_5 Z^{-5/17} + A_{12} + A_{13}T_r^{w} + A_{14}(T_r^{w})^2$$
$$+ A_{15}(a_6 - T_r^{w})^{10} + A_{16}(a_7 + (T_r^{w})^{19})^{-1}$$
$$- (a_8 + (T_r^{w})^{11})^{-1} (A_{17} + 2A_{18}P_r^{w} + 3A_{19} (P_r^{w})^2)$$
$$- A_{20}(T_r^{w})^{18} (a_9 + (T_r^{w})^2) (-3 (a_{10} + P_r^{w})^{-4} + a_{11})$$
$$3A_{21} (a_{12} - T_r^{w}) (P_r^{w})^2 + 4A_{22} (T_r^{w})^{-20} (P_r^{w})^3\}^{-1}, \quad (19)$$

where $a_1,\ldots,a_{12}$, $A_{11},\ldots,A_{22}$ are constants for water density function, ρ_1 is water density, P_r^{w} is water reduced pressure, T_r^{w} is water reduced temperature, v_c^{w} is water critical specific volume.

2.39 Oil viscosity

The oil viscosity relation with temperature (Reid *et al.* 1987; White & Oostrom 2000) is given by,

$$\mu^{o} = \exp(A + B/T) \quad (20)$$

where μ^{o} is oil viscosity A and B are oil viscosity constants

2.40 Oil density

For oil density we use properties of perchloroethylene (PCE) (Reid *et al.* 1987). Oil density then as a function of temperature is determined using equation (21-23) i.e. Modified-Rackett method under saturated conditions,

$$\rho_{sat}^{\ o} = \{[R^{o}T_c^{\ o} / P_c^{\ o}](ZRA^{o})^{(1+(1-T_r^{\ o})^{2/7})}\}^{-1} \quad (21)$$

$$T_r^{\ o} = T / T_c^{\ o} \quad (22)$$

$$Z_{RA}^{\ o} = 0.29056 - 0.08775 w_p^{\ o} \quad (23)$$

where $\rho_{sat}^{\ o}$ is the oil density under saturated condition, R^{o} is specific ideal gas constant for oil, $T_r^{\ o}$ is the oil reduced temperature, $T_c^{\ o}$ is the critical temperature for oil, $P_c^{\ o}$ is the critical pressure for oil, $w_p^{\ o}$ is the Pitzner accentric factor for PCE, $Z_{RA}^{\ o}$ is the Rackett compressibility factor.

2.4 Calculation of Dynamic/Damping coefficient

Damping coefficient (τ) can be determined by using equation (2), which in turn requires the estimation of average capillary pressure for the domain. The averaging procedures that have been employed in this work for the calculation of (τ) can be referred to in the works of Das *et al.* (2007), Mirzaei & Das (2007).

3.0 Computational Results & Discussions

3.1 Simulator Used, Grid Refinement & Convergence

The two phase flow i.e. "water-oil" simulations in this work have been carried out using the "*Subsurface Transport Over Multiple Phases*" (STOMP) (http://stomp.pnl.gov/) simulator (White & Oostrom, 2006) written in FORTRAN. The code has been previously validated against a range of engineering problems which include enhanced oil recovery operations (Scroth *et al.* 1998; Ataie-Ashtiani *et al.* 2001 & 2002; Das *et al.* 2004). The governing equations of the two phase porous flow model are discretized using the Finite Volume technique whilst the temporal discretization procedure follows the Implicit-Theta time-stepping approach. Simulation results were obtained on a computational grid which was refined by increasing the total number of cells by 25%, and no major differences were observed. Convergence was achieved by altering the maximum Newton-Rhapson iteration between 16 and 50, and by applying a tolerance value of 10^{-6}. Results indicate that grid refinement has negligible effect on the final capillary pressure-saturation curves. A maximum time step of 600s and time acceleration factor 1.125 have been chosen to yield stable solutions.

3.2 Computational Domain & Description of Test Cases

A 3-D cylindrical column shown in Figure 1 similar to the laboratory micro-scale set-ups (Mirzaei and Das, 2006) have been used in this work.

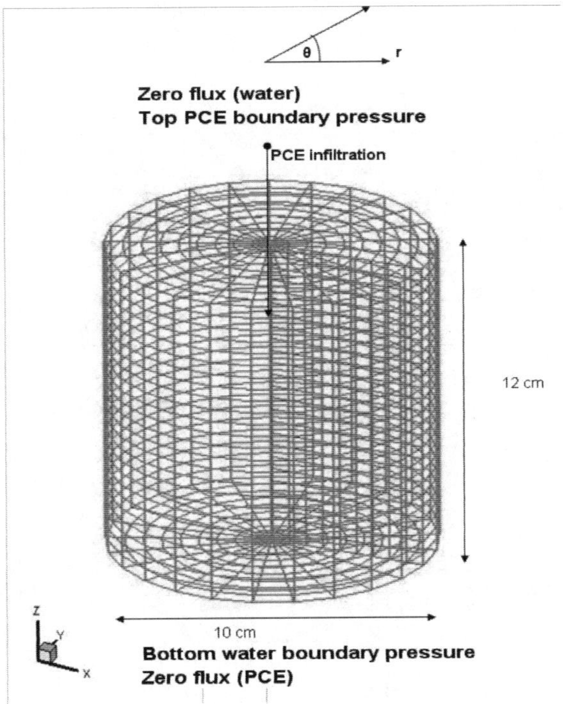

Figure 1: 3-D Cylindrical Domain for homogenous porous media

The diameter of the column is 10 cm whilst the vertical length in Z-direction is 12 cm. The computational grid consists of 4x4x24 cells with 416 nodes. The nodal spacings for the domain are listed in Table 1. As heterogeneity effects have not been included the domains are fully isotropic.

Table 1: Nodal spacing for the Cylindrical Domain

Domain geometry	Number of nodes x nodal spacing		
	N x Δ r (cm)	N x $\Delta\theta$ (degree)	N x Δz (cm)
3D Cylindrical	4 x 1.25	4 x 90	1 x 0.05
			24 x 0.5
			1 x 0.05

Table 2 lists the simulations that have been carried out at different temperatures for both coarse and fine sands. At any given temperature, four dynamic simulations and one steady-state simulation has been carried out with a view to generate numerically stable and reliable capillary pressure curves.

Table 2: Simulation Test Cases

Simulation Case	Temperature (°C)			
Dynamic_1	20	40	60	80
Dynamic_2	20	40	60	80
Dynamic_3	20	40	60	80
Dynamic_4	20	40	60	80
Steady-State	20	40	60	80

3.4 Material & Physical parameters

The material properties used in this work correspond to two different sands (coarse & fine sands) and are listed in Table 3. It should be noted that the material properties do not vary with temperature.

Table 3: Material Parameters for Sands

Properties	Coarse Sand	Fine Sand
Permeability, K (m^2)	5×10^{-9}	5×10^{-12}
Porosity, ø (-)	0.4	0.4
Entry Pressure, Pd (Nm^{-2})	375	1325
Pore size distribution index, λ (-)	3.86	2.49
Bulk Density, ρ (kg m-3)	2630	2650

The various temperature dependent fluid physical parameters based on the empirical relations (15-23) are listed in Table 4. Simulations have been carried out in the range of 20°C-80°C and the effects of temperature have also been included on the initial and bottom boundary pressures.

Table 4: Temperature Dependent Fluid Physical Parameters

Fluid Properties	Temperature (°C)			
	20	40	60	80
Residual saturation, S_{wr}	1.0000×10^{-2}	9.7600×10^{-2}	1.8520×10^{-1}	2.7280×10^{-1}
Water-air surface tension, γ^{lg} dynes/(cmK)	7.2759×10^{1}	6.9612×10^{1}	6.6249×10^{1}	6.2684×10^{1}
Water-Oil interfacial tension, γ^{lo} dynes/(cmK)	4.4987×10^{1}	4.4147×10^{1}	4.3308×10^{1}	4.2468×10^{1}
Water density, ρ^{l} kg/m^3	9.9828×10^{2}	9.9227×10^{2}	9.8314×10^{2}	9.7163×10^{2}
PCE density, ρsat^{o}	1.6668×10^{3}	1.6348×10^{3}	1.6020×10^{3}	1.5682×10^{3}
Water viscosity, $\mu^{l} \times 10^{-3}$ Pa.s	1.0177×10^{1}	6.6800×10^{-1}	4.7648×10^{-1}	3.6247×10^{-1}
PCE viscosity, $\mu^{o} \times 10^{-3}$ Pa.s	8.9977×10^{-1}	7.3212×10^{-1}	6.1065×10^{-1}	5.1990×10^{-1}
Water initial presure, P_{wip} Pa	102473.77	1.02476.85	102456.34	102443.10
Water bottom pressure, P_{wbp} Pa	102507.99	102500.87	102490.04	102476.40

3.5 *Initial and Boundary Conditions*

Capillary pressure-saturation relationships and the dynamic coefficient can be determined by simulating the pressure cell experiment for drainage process and can be referred to in the works of Ataie-Ashtiani *et al* (2002) and Mirzaei & Das (2006). At the top face of the porous domain we inject PCE at different pressures for different dynamic cases listed in Table 5.

Table 5: Boundary pressure for dynamic and steady-state simulation at 20°C

Simulation Test Case	**Top boundary**	**Bottom boundary**			
	20^0C-80^0C	20^0C	40^0C	60^0C	80^0C
	Dirichlet NAPL pressure (Pa)	Zero flux NAPL			
	Zero flux water	Dirichlet water pressure (Pa)			
Dynamic_1	109316	102507.9869	102500.8658	102490.0413	102476.4032
Dynamic_2	112513	102507.9869	102500.8658	102490.0413	102476.4032
Dynamic_3	117308	102507.9869	102500.8658	102490.0413	102476.4032
Dynamic_4	122903	102507.9869	102500.8658	102490.0413	102476.4032

3.6 *Quasi-Static & Dynamic Simulation Procedure*

a) Quasi-static simulations are carried out by initializing the PCE pressure to zero value. Subsequently, the PCE pressure at the top of domain is gradually increased until the capillary pressure (P^c) equates to the porous domain entry pressure (P^d) when PCE starts to infiltrate the porous domain. Simulations are carried out until saturation at each grid point does not change with time or the variation is smaller than the tolerance limit (10^{-6}). The simulation is continued by increasing the PCE pressure at the top for the attained P^c-S curves. Simulation is stopped when the residual saturation is attained at P^c equal to 11000 Pa.

b) Dynamic simulations are carried out by imposing the high PCE pressure at the top of the domain for every simulation until the residual saturation is attained.

c) To carry out the final integration, an in-house FORTRAN program which performs volume averaging to determine saturation weighted-average of capillary pressure is used. For determining the dynamic coefficient knowledge of saturation time derivative data and the capillary pressure data under dynamic and steady-state conditions is processed from the simulation output. Averaged data for ($P^{c,dyn}$- $P^{c,equ}$) and ($\partial S/\partial t$) are then fitted on to the straight line by carrying out linear regression analysis. If the R^2 values are in the vicinity of 1 ($R^2 \geq 0.95$), it can be assumed that the data fits well on the straight line. The final capillary pressure-saturation curves are then generated using Microsoft Excel. As the set of output data is huge, Tecplot Macro files which can relatively reduce the times for post-processing is generated.

3.7 *Dynamic & Steady State Capillary Pressure Saturation Curves*

Figures 2 & 3 illustrate the dynamic and steady state capillary pressure saturation curves for coarse and fine sands at 20^0C. In Figure 2 for coarse sand at 20°C, initial capillary pressure is 224 Pa. At high saturation between 1 and 0.35, the capillary pressure of dynamic cases (1-4) gradually rises as the saturation increases. However, there is a sudden pressure drop in the steady-state case from 270-249 Pa between saturation value of 0.449 and 0.416. This occurs in the very start of drainage process between 6s and 2 hours. At saturation value of 0.325 a pressure jump for all dynamic cases is observed, at 60s in dynamic case 4 (PCE-head 135cm) and at 1200s in dynamic case 1 (PCE-head 50cm). After this jump, the capillary pressure further increases as the saturation decreases.

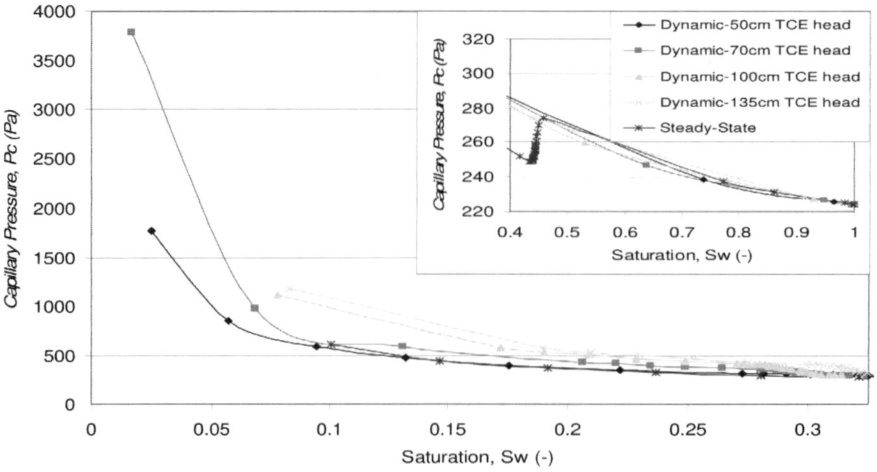

Figure 2: Dynamic & Steady-State P^c-S Curve for Coarse Sand @ 20^0C

Figure 3 represents the P^c-S curves for fine sand at 20°C. Initial capillary pressure 800 Pa in this case is higher than for coarse sand at the same temperature. No pressure jump is observed as in the case of course sand. In the dynamic and the steady-state cases stable increase in P^c is observed from 1200s to 2 hours. At lower saturation values, from 0.4 to residual saturation, P^c increases rapidly whilst the saturation decreases much slower. This occurs after 9-10 hours. The increment of P^c in each dynamic case is not as drastic as the increment in coarse sand.

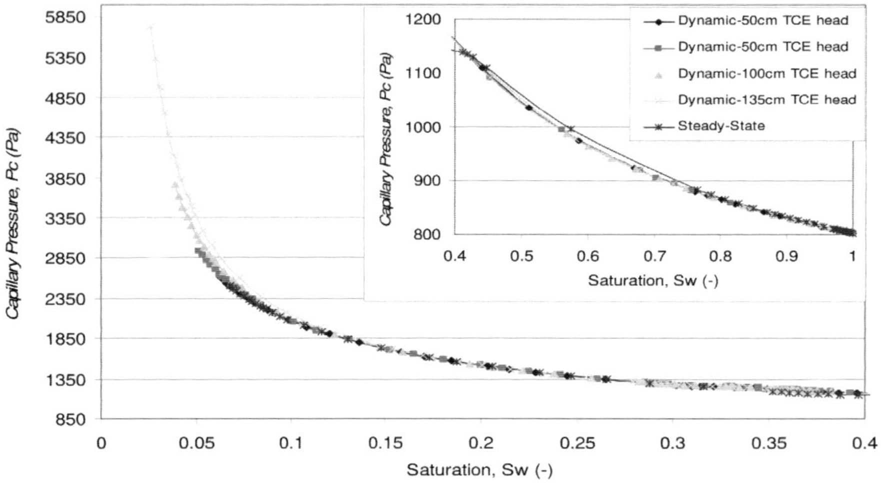

Figure 3: Dynamic & Steady-State P^c-S Curve for Fine Sand @ 20^0C

The pressure jump observed in the case of coarse sand is also known as the Haines jump (Haines 1930) which occurs because of the presence of viscous or gravity forces along with the low entry pore space pressure and higher intrinsic permeability. This results in a jump when the PCE pressure outcomes the entry pressure of pore space and starts to displace the entrapped water inside each pore space. This also explains the fact that in the case of fine sand where the intrinsic relative permeability is much lower and the entry pore space pressure is higher, Haines jump does not occur.

In general, the calculated dynamic coefficients at 20^0C (Table 6) are higher for fine sands as in comparison to the coarse sands. Elevated temperatures also cause a higher capillary pressure and a smaller saturation time derivative. Dynamic coefficient at higher temperatures is higher than its values at lower temperature for the same water saturation. These trends are more significant after a critical saturation (low saturations near the residual saturation) at each temperature. The R^2 values show that the data in some cases may not be linear. This is more significant in the case of fine sand domain where the displacement process is very slow and the effects of difference in injection pressure are not as significant as in the coarse sand. The values of dynamic coefficients in fine sand for different temperature range between $6.4x10^6$-$1.1x10^{11}$ Pa.s whilst in the case of coarse sand the range is between $4.2x10^6$-$5.33x10^8$ Pa.s.

Table 6: Dynamic Coefficients for Coarse and Fine Sands at 20°C

Coarse Sand 20°C			Fine Sand 20°C		
S_w	τ (Pa.s)	R^2	S_w	τ (Pa.s)	R^2
$2.80x10^{-1}$	$4.20x10^6$	0.781	$3.10x10^{-1}$	$6.40x10^6$	0.99
$2.73x10^{-1}$	$4.30x10^6$	0.98	$2.98x10^{-1}$	$1.00x10^7$	0.96
$2.37x10^{-1}$	$6.00x10^6$	0.98	$2.04x10^{-1}$	$1.60x10^7$	0.7
$2.00x10^{-1}$	$8.90 x10^6$	0.99	$9.84x10^{-2}$	$2.10x10^9$	0.62
$1.91x10^{-1}$	$9.10x10^6$	0.98	$7.14x10^{-2}$	$3.50x10^9$	0.8

Tables 7 and 8 correspond to the times required for attaining the residual saturation for coarse and fine sands at 20^{0}C.

Table 7: Times required for attaining residual saturation for coarse sand at 20°C

20°C		Coarse Sand	
	Time (hr)	**Final Saturation**	**Residual Saturation**
Dynamic_1	8	0.00999	0.00999
Dynamic_2	5	0.00999	0.00999
Dynamic_3	3	0.00999	0.00999
Dynamic_4	3	0.00999	0.00999
Steady-State	9	0.101	0.00999

Table 8: Times required for attaining residual saturation for fine sand at 20°C

20°C		Fine Sand	
	Time (hr)	**Final Saturation**	**Residual Saturation**
Dynamic_1	975	0.458	0.00999
Dynamic_2	975	0.0337	0.00999
Dynamic_3	975	0.0193	0.00999
Dynamic_4	975	0.0103	0.00999
Steady-State	975	0.0488	0.00999

3-D saturation plots presented in Figures 4(a-f) and 5(a-f) illustrate the times required to attain the residual saturation at different temperatures for coarse and fine sands under dynamic conditions at 20^{0}C and 80^{0}C respectively. It can be inferred from Figures 4(a-f) and 5(a-f), that the effects of temperature are visible at earlier time steps i.e. at 30s, 60s until 1 hour, when sudden displacements take place.

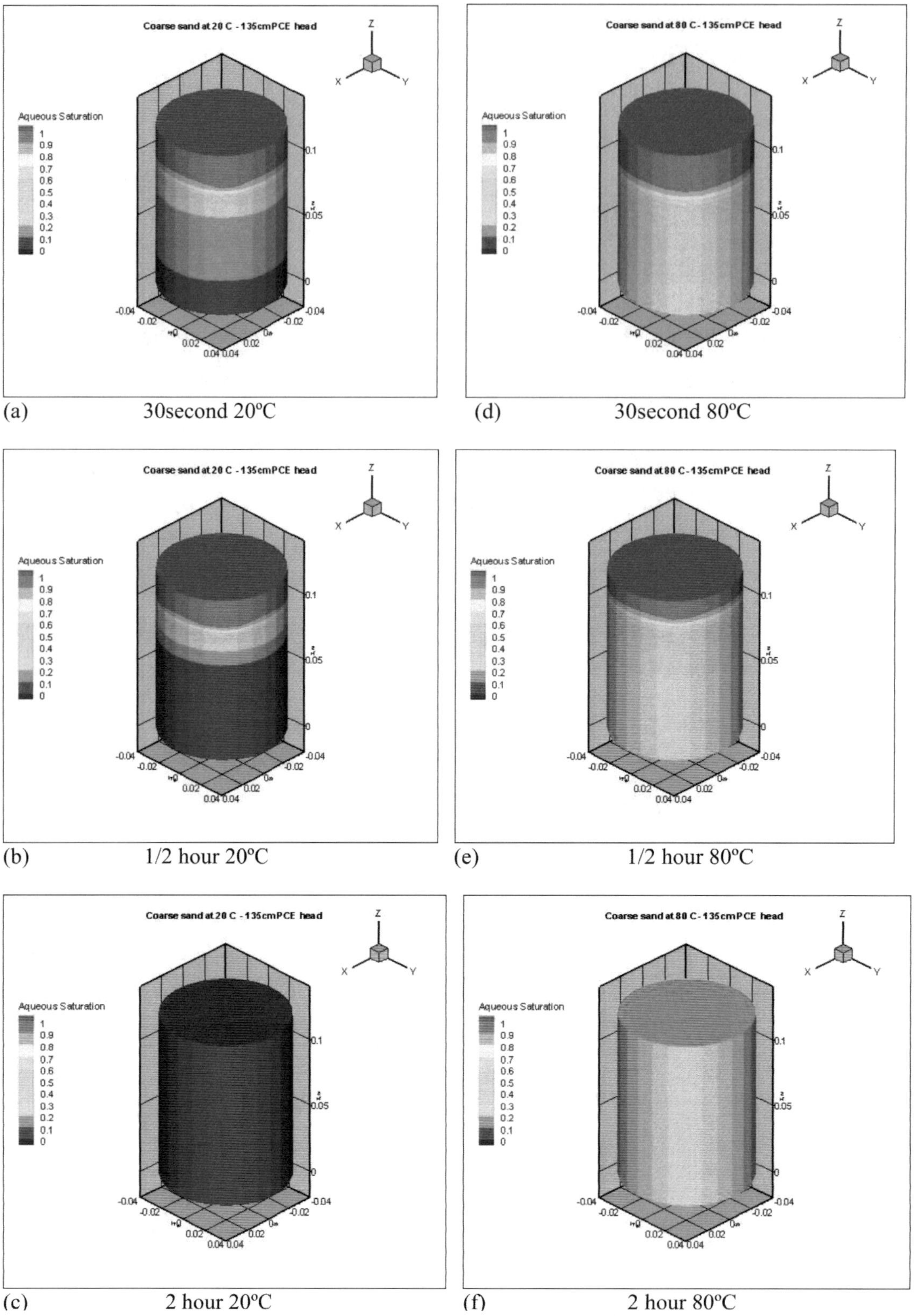

Figure 4: Aqueous saturation in Coarse Sands at 20°C(a-c) & 80°C (d-f) for 2 hours

Then, the displacement gets slower which is marked by a slower decrease in water saturation. In 20°C case at 3 s and 60s, PCE displaces a lot of water at the bottom of the porous domain.

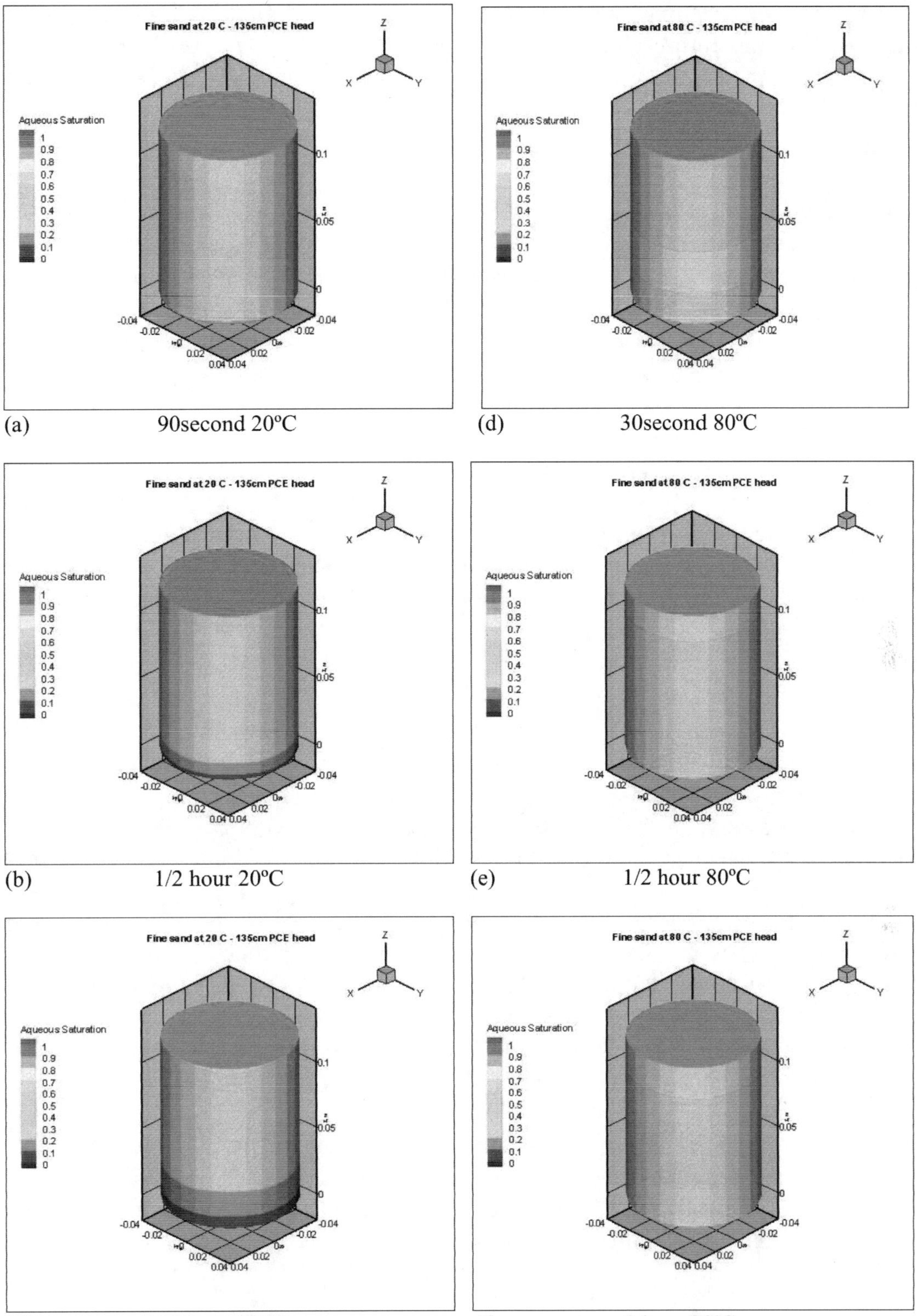

(a) 90second 20ºC (d) 30second 80ºC

(b) 1/2 hour 20ºC (e) 1/2 hour 80ºC

(c) 2 hour 20ºC (f) 2 hour 80ºC

Figure 16: Aqueous saturation in fine sand at 20ºC (a-c) and 80ºC (d-f) for 2 hour

When the temperature is increased, water amount that is displaced in the first 30s and 60s is decreased. At 80ºC, water amount that is displaced in the first 30s and 60s is reduced. In addition, an effect of gravity and viscous forces along with low water pressure may play an important role to slow the displacement as the temperature increases. The water displacement is not distributed evenly at each node in the cylinder domain. First, the water content in the bottom is displaced, then the displacement is going upward with water in the side of domain is displaced earlier than water in the middle of the domain.

3.0 Conclusions

Simulations have been carried out using STOMP 4.0 to quantify the effects of temperature and fluid properties on the dynamic capillary pressure and in turn on dynamic coefficients for two-phase porous flow in the cylindrical domain. Series of runs were carried out within the range of 20^0C-80^0C for two different materials (coarse and fine sands) under dynamic conditions to determine the values of dynamic coefficients. The dynamic coefficients found in this work are non linear function of saturation and temperature. This is contrary to the result by Stauffer (1978) which suggests that the dynamic coefficient is independent of saturation. Our results are consistent with most of the recent studies by (Hassanizadeh et al, 2002; Manthey et al, 2005; O'Carroll et al, 2005; Mirzaei and Das, 2006), however the temperature effects as reported in this work have not been reported before. We observe that with an increase in temperature the dynamic coefficients increase and at any given saturation value large time scales are required for non-aqueous phase flow (NAPL) displacements.

4.0 References

Ataie-Ashtiani, B., Hassanizadeh, S.M., Oostrom, M., Celia, M.A., White, M.D., 2001. Effective parameters for two-phase flow in a porous medium with periodic heterogeneities. Journal of Contaminant Hydrology 49, 87-109.

Ataie-Ashtiani, B., Hassanizadeh, S.M., Celia, M.A., 2002. Effects of heterogeneities on capillary pressure-saturation relative permeability relationships. Journal of Contaminant Hydrology 56, 175-192.

Ataie-Ashtiani, B. Hassanizadeh, S.M., Oung, O., Westrate, F.A., Bezuijen, A., 2003. Numerical modeling of two-phase flow in geocentrifuge. Environmental Modelling & Software 18, pp 231-241.

Bear, J., Verruijt, A.1972. Modelling Groundwater Flow & Pollution, Theory and Applications of Transport in Porous Media. Boston, MA: Reidel, 1987

Beliaev, A.Y., Hassanizadeh, S.M., 2001. A theoretical model of hysterisis and dynamic effects in the capillary relation for two-phase flow in porous media. Journal of Transport in Porous Media 43, 487-510.

Beliaev, A.Y., Schotting, R.J., 2001. Analysis of a new model for unsaturated flow in porous media including hysteresis and dynamic effects. Computational Geosciences 5, 345-368.

Berentsen, C.W.J., Hassanizadeh, S.M., Bezuijen, A., Oung, O., 2006. Modelling of two-phase flow in porous media including non-equilibrium capillary pressure effects. In: Proceedings of the XVI International Conference on Computational Methods in Water Resources (CMWR), Copenhagen, Denmark

Blunt MJ, King P. Relative Permeabilities from 2-dimensional and 3-dimensional pore scale network modelling. Transport in Porous Media, 1991; 6(4):407-33.

Brooks, R.H., Corey, A.T., 1964. Hydraulic Properties of Porous Media. Hydrology Papers, Colorado State University

Dahle, H.K., Celia, M.A., Hassanizadeh, S.M., 2005. Bundle of tubes model for calculating dynamics effects on upscaled properties of two-phase flow in porous media. Transport in porous Mediqa 56, 329-350

Das, D.B., Hassanizadeh, S.M., 2005. Upscaling Multiphase Flow in Porous Media: From Pore to Core and Beyond. Springer, Berlin pp 1-3, ISBN: 1-4020-3513-6

Das, D.B., Hassanizadeh, S.M., Rotter, B.E, Ataie-Ashtiani, B., 2004. A numerical study of micro-heterogeneity effects on upscaled properties of twophase flow in porous media. Transport in Porous Media 56, 329-350.

Das, D.B., Mirzaei, M., Widdows, N., 2006. Non-uniqueness in capillary pressure-saturation-relative permeability relationship for two phase flow in porous media: implications of intensity and random scale micro-heterogeneity. Journal of Chemical Engineering Science 61, 6786-6803.

Das, D.B., Gauldie, R., Mirzaei, M., 2007. Dynamic effects in capillary pressure relationships for two-phase flow in porous media: Implications of fluid properties. AIChe Journal Manuscript.

Davis, E.L., 1994. Effect of temperature and pore size on the hydraulic properties and flow of a hydrocarbon oil in the subsurface. Journal of Contaminant Hydrology 16, 55-86.

Demond, A.H., Roberts, P.V., 1987. An examination of relative permeability relations for two-phase flow in porous media. American Water Resources Bulletin.

Farouki, T.O., Thermal properties of soil, 1986. Trans Tech Publications, Clausthal-Zellerfeld, Germany, pp 29-30.

Gatmiri, B., Delage, P.,1997. A formulation of fully coupled thermal hydraulic- mechanical behaviour of saturated porous media – Numerical approach. International Journal for Numerical and Analytical Methods in Geomechanics 21, 199-225.

Geilen A.J.M., Hassanizadeh, S.M., Nordhaug, H.F., Leijmse A., 2005. Dynamic effects in multiphase flow: a pore-scale network approach. In: Das, DB., Hassanizadeh, S.M. (Eds.), Special Issue on: Upscaling Multiphase Flow in Porous Media: From Pore to Core and Beyond. Springer, The Netherlands

Grant, S.A, 2003. Extension of a temperature effects model for capillary pressure-saturation relations. Water Recources Research 39, no 1, 1003, doi:10.1029/2000WR000193.

Grant, S.A, Bachmann, J. Effect on temperature on capillary pressure.

Grant, S.A., Salehzadeh, A., 1996. Calculation of temperature effects on wetting coefficients of porous solids and their capillary pressure functions.Water Resources Research 32, no 2 pp 261-270

Haines, W.B., 1930. Studies in the physical properties of soil hysteresis effect in capillary properties and the modes of moisture distribution associated therewith. Journal of Agricultural Science 20, 97-116

Hanyga, A., Jianfei, L., 2004. Thermal effects in immiscible two-fluid porous flow. International Journal of Engineering Science 42, 291-301.

Hanyga, A., Seredynska, M., 2005. A dynamic model of capillary hysteresis in immiscible fluid displacement. Transport in Porous Media 59, 294-265.

Hassanizadeh, S.M., Gray, W.G., 1990. Mechanics and thermodynamics of multiphase flow in porous media including interphase boundaries. Adv. Water Resources 13, no 4.

Hassanizadeh, S.M., Gray, W.G., 1993 Thermodynamic basic of capillary pressure in porous media. Water Resources Research 29 (10), 3389-3405.

Hassanizadeh, S.M., Celia, M.A., Dahle, H.K., 2002. Dynamic effect in the capillary pressure-saturation relationship and its impact on unsaturated flow. Agricultural Sciences 7(2), 69-71

Helmig, R., 1997. Multiphase flow and transport processes in the subsurface. Springer, Berlin. 367pp.

Imhoff, P.T., Frizzell, A., Miller, C.T, 1997. Evaluation of thermal effect on the dissolution of nonaquaeus phase liquid in porous media. Environmental Science and Technology 31, 1615-1622.

Kueper, B.H., Frind, E.O., 1989. An overview of immiscible fingering in porous media. Journal of Contaminant Hydrology 5, 83–95.

Kozak, J.A., Ahuja, L.R., 2005. Scaling of infiltration and redistribution of water accros soil texture. Soil Science Society.

Kovscek, A.R., 2003. Heavy and Thermal Oil Rcovery Production Mechanisms. Quarterly Technical Progress Report, DE-FC26-00BC15311.

Mandal, J.N., Divshikar, D.G, 1995. Soil testing in civil engineering. A.A Balkema Publisher, Brookfield, USA, pp.10-90.

Manthey, S., Hassanizadeh, S.M., 2005. Macro-scale dynamic effects in homogeneous and heterogeneous porous madia. Transport in Porous Media 58, 121-145.

Marshall, T.J., Holmes, J.W., 1973. Soil Physics. Cambridge University Press

Meyer, C. A., Mclintock, R. B., Silvestri, G. J., Spencer Jr, R. C., 1967. ASME steam tables: Thermodynamic and transport for steam and water, pp 13-30.

Mirzaei, M., Das, D.B., 2007. Dynamic effects in capillary pressure saturations relationships for two-phase flow in 3D porous media: Implications of micro-heterogeneities. Chemical Engineering Science 62, 1927-1947.

Muralidhar, K., Sheorey, T., 2003. Isothermal and non-isothermal oil-water flow and viscous fingering in porous media. International Journal of Thermal Sciences 42, 665-676.102

Murdoch, A.I., Haassanizadeh, S.M., 2002. Macro-scale balance relations for bulk, interfacial and common line systems in multiphase flows through porous media on the basis of molecular considerations. Internationa Journal of Multiphase Flow 28, 1091-1123

Muskat, M., Mears, M., 1936. The flow of heterogeneous fluids through porous media. Physics 7, 346–363.

Narasimhan, A., Lage, J.L., 2001. Modified Hazen-Dupuit-Darcy model for forced convection of a fluid with temperature-dependent viscosity. Journal of Heat Transfer 123.

Nordhaug HF, Celia M, Dahle HK. A pore network model for calculation of interfacial velocities. Advances in Water Resources 2003; 26(10):1061-1074.

Nichols, W.E., Aimo, N.J., Oostrom, M., White, M.D., 1997. STOMP application guide. PNNL-11216, UC-2010.

Nutt, C.W., 1981. The physical basis of the displacement of oil from porous media by other fluids: A capillary bundle model. Proceeding of The Royal Society of London. Series A, Mathematical and Physical Science 382, pp. 155-178.

O'Carroll, D.M., Phelan, T.J., Abriola, L.M., 2005. Exploring dynamic effects in capillary pressure in multistep outflow experiments. Water Resources Research 41, W11419doi: 10.1029/2005WR004010.

Oostrom, M., Meck, D.H., White, M.D., 2003. STOMP an introductory short course. Pacific Northwest National Laboratory, Richland, Washington.

Oung, O., Hassanizadfeh, S.M., Bezuijen, A., 2005. Two-phase flow experiments in a geocentrifuge and the significance of dynamic capillary pressure effect. Journal of Porous Media 8(3), 247-257

Panfilov M, Panfilova I. Phenomenological Meniscus Model for Two-Phase Flows in Porous Media. Transport in Porous Media 2005; 58(1-2):87-119.

Reid, R. C., Prausnitz, J. M., Poling, B. E., 1987. The Properties of gas and liquids. Mcgraw Hill.Inc, pp 433-485.

Schembre, J.M., Kovscek, A.R., 2005. Mechanism of formation damage at elevated temperature. Journal of Energy Resources 127.

Schembre, J.M, Kovscek, A.R., 2006. Estimation of dynamic relative permeability and capillary pressure from countercurrent imbibition experiments. Transport in Porous Media 65: 31-51.

Schembre, J.M., Tang, G.-Q., Kovscek, A.R., 2006. Interrelationship of temperature and wettability on the relative permeability of heavy oil in Diatomaceous Rocks. SPE 938831 (Society of Petroleum Engineers).

Schroth, M.H., Istok, J.D., Selker, J.S., Oostrom, M., White, M.D., 1998. Multifluid flow in bedded porous media: laboratory experiments and numerical simulations. Advances in Water Research 22 (2), 169–183

She, H.Y., Sleep, B.E., 1998. The effect of temperature on capillary pressure saturation relationships for air-water and perchloroethylene-water systems. Water Resources Research 34, no10 pp2587-2597.

Sinnokrot, A.A., Ramey Jr, H.J., Marsden Jr, S.S., 1971. Effect of temperature level upon capillary pressure curves. Society of Petroleum Engineers (SPE) Journal 11, pp13-22.103

Stauffer F., 1979 Time dependence of the relationship between capillary pressure, water content and conductivity during drainage of porous media. In proceedings of the IAHR conference on scale effects in porous media, Thessaloniki, Greece.

Theodoropoulou, M.A., Sygouni, V., Karoutsos, V., Tsakiroglou, CD 2005. Relative permeability and capillary pressure functions of porous media as related to the displacement growth pattern. International Journal of Multiphase Flow 31 (10-11), 1155-1180

Tzakiroglou CD, Theodoropoulou MA, Karoutsos V. Nonequilibrium Capillary Pressure and Relative Permeability Curves of Porous Media. AIChE Journal 2003, 49(10):2472-2486.

White, M.D., Oostrom, M., 2000. STOMP 2.0 theory guide. PNNL-12030, UC-2010.

White, M.D., Oostrom, M., 2006. STOMP 4.0 user guide. PNNL-17782.

New insights into the reaction of citric acid with calcite

M.H. Alkhaldi, H.A. Nasr-El-Din,* H. Sarma, University of Adelaide,
* Texas A&M University

Abstract

The previous studies of citric-calcite have been limited to core flood tests, and bench scale experiments. However, the reaction kinetics of citric acid-calcite has not been measured. This paper gives, for the first time, the kinetics of citric-calcite reaction, which will provide a better way to model the performance of citric acid as a stand-alone stimulation fluid.

In this paper, the rotating disk was used to study citric-calcite reaction at pressure of 1,000 psi, temperature values from 25 to 50°C, citric acid concentrations of 1 to 7.5 wt%, and disk rotational speeds of 100 to 1,000 rpm. The overall reaction rate was mainly governed by the rate of the forward reaction, and hence, it was molded by a new simplified reaction rate equation, rate = $k_f \{K_{a1} * C_B\}^{n/2}$. The average value of the reaction order (n) was found to be equal to **0.833**.

Introduction

Hydrochloric acid (HCl) has been widely used for stimulating carbonate formations, limestone and dolomite, in both matrix and acid fracturing treatments. However, its reaction rate with carbonate minerals is very rapid, which results in poor etched fracture lengths and high corrosion rates, especially at high temperature formations. Therefore, acid systems such as organic acids have been used as alternatives to HCl to improve the efficiency of the stimulation treatments (Harris, 1961; Chatelain *et al.*, 1976; Crowe *et al.*, 1988; Fredd and Fogler, 1998a; Fredd and Fogler, 1998b; Fredd and Fogler, 1998c; Fredd, 2000; Huang *et al.*, 2000; Nasr-El-Din *et al.*, 2001; Buijse *et al.*, 2004; Nasr-El-Din *et al.*, 2006; Al-Mohammad *et al.*, 2006; Al- Al-Mutairi *et al.*, 2007; Nasr-El-Din *et al.*, 2008; Mutairi *et al.*, 2008).

In order to overcome the concerns of HCl acid, organic acids are usually used as an alternative to HCl in high temperature formations. The main organic acids that are frequently used in stimulation treatments are acetic, formic, and citric acid. These acids are less reactive with carbonate rocks than hydrochloric acid. The retarded nature of these acids makes them a good alternative to HCl acid. However, organic acids are more expensive than HCl per unit volume of rock dissolved. They cannot be used at high acid concentrations. Typically, the concentration of acetic and formic acids used is < 13 wt% and < 9 wt%, respectively. This is because the reaction products (especially calcium formate) can precipitate at high acid concentrations (Robert and Crowe, 2000). In addition, the reaction of organic acids with calcite is reversible and thermodynamically limited by the presence of the reaction products. In other words, the reaction is controlled by the diffusion of the reaction products away from the rock surface (Fredd, 2000; Fredd and Fogler, 1998c).

Recently, citric acid has been used in some field treatments as an alternative to other weak (organic) acids, formic acid and acetic acid. It was introduced because it can be used in an encapsulated form, which prevents its reaction with both the production tubular and the formation until 180°F (Blauch *et al.*, 2003; Burgos *et al.*, 2004), and hence this will save the cost of the use of the corrosion inhibitors. This use of encapsulated citric acid in petroleum fields has met with mixed field results. For example, Burgos *et al.* (2004) reported that there was no improvement in the performance of an acid-fractured well after the application of encapsulated citric acid. They explained the unexpected field results in terms of calcium citrate precipitation, which is one of the main products of the citric acid reaction with calcite. However, Blauch *et al.* (2003) reported positive field results and that the formation damage because of calcium citrate precipitation is not a major concern at down-hole conditions. These mixed results of citric acid applications in field treatments indicate the need for an improved understanding of citric acid reactivity with calcite.

Two factors affect the rate of the reactivity of citric acid with calcite: the mass transfer rate and the surface reaction rate. The mass transfer rate is a measure of the rate at which citric acid is transported from the bulk solution to the surface of the reaction. The reaction rate is a measure of the rate at which the citric acid reaction with calcite proceeds at the surface of the reaction.

The parameters affecting the citric acid mass transfer have been extensively studied by Alkhaldi *et al.* (2007). They investigated the reaction of citric acid with calcite using the rotating disk apparatus. They examined the effects of the disk rotational speed, the system pressure, and the presence of magnesium and iron ions on the citric-calcite reaction. Using initial citric acid concentration of 7.5 wt%, they found that the reaction of citric acid with calcite is mass-transfer limited up to 500 rpm at 50°C. They also reported that the precipitation of calcium citrate on the surface of the rotating disk limited the reaction of citric acid with the calcite disk.

The reaction kinetics of HCl and several organic acids such as acetic has been extensively studied (Lund *et al.*, 1975; Plummer, 1978; Fredd and Fogler, 1998b; Fredd and Fogler, 1998c; AlKattan *et al.*, 1998; AlKattan *et al.*, 2002; Buijse *et al.*, 2004; Taylor *et al.*, 2004; Taylor *et al.*, 2006). These studies provide the reaction kinetics parameters, which are essential in designing different stimulation treatments. In addition, these parameters are used as inputs in commercial simulation software to optimize stimulation treatments. The reaction kinetics of citric acid with calcite has not been studied in detail before. Therefore, the present study uses the rotating disk instrument to study the reaction kinetics of citric acid reaction with calcite. The objectives of this study are: (1) develop a simple equation to model the reaction of citric acid with calcite, (2) determine the parameters of the reaction kinetics of citric acid reaction with calcite, (reaction order, and reaction constant), and (3) study the effect of temperature on the reaction constant.

Acid Reaction Rate

Previous studies have shown that the rate of the reaction of HCl with calcite can be described using the power-law model as follows (Lund *et al.*, 1975; AlKattan *et al.*, 1998; AlKattan *et al.* 2002; Taylor *et al.*, 2004; Taylor *et al.*, 2006):

$$R = k\,([H^+]_s)^n = k\,([HCl])^n \qquad (1)$$

where R is the surface reaction rate (mol/cm^2/s), k is the reaction rate constant, $[H^+]_s$ is the concentration of H^+ at the reacting surface, and n is the reaction order. Calcite dissolution in HCl solutions is limited by the mass transfer rate. Therefore, the overall HCl-carbonate reaction rate is expressed as follows (Buijse *et al.*, 2004):

$$\text{calcite dissolution} = k_{mt}\,[H^+]_b = k_{mt}\,[HCl] \qquad (2)$$

where k_{mt} is the mass transfer coefficient, and $[H^+]_b$ is the concentration of H^+ in the bulk solution.

The model described by Eqs. (1) and (2) is not applicable for weak organic acids. Several studies have shown that this model fails when applied to organic acids such as acetic and formic (Fredd and Fogler, 1998c, Buijse *et al.*, 2004). This is mainly due to the reversible nature of these acids. Buijse *et al.* (2004) showed that the reaction of these acids with calcite can be expressed using Eq. 3:

$$R = k_f \left\{ [H^+]_s^n - X^n \cdot \frac{(C_0 - C_S)^{2n}}{[H^+]_s^n} \right\} \qquad (3)$$

where R is the reaction rate, k_f is the reaction constant of the forward reaction, C_0 is the initial acid concentration, C_s is the acid concentration at the reacting surface, n is the reaction order, and X is a function of carbonate species equilibrium constants and the solubility product of calcium carbonate, and its value is equal to 2 E-10 (Buijse *et al.*, 2004).

A new simplified version of Buijse *et al.*'s model is used to study the reaction of citric acid with calcite. The simplified model was derived based on the following assumptions:

1. The reversible reactions are partially suppressed due to calcium citrate precipitation. As calcium citrate precipitates, calcium ions are depleted at the surface and hence the equilibrium reaction is shifted toward the forward reaction. In addition, the reduction of the free di-hydrogen citrate anion will further shift the equilibrium reaction shown towards the forward direction, (toward the products), and may increase the spending of citric acid, (Chang *et al.*, 2008). Therefore, it is assumed that the effect of the reversible reactions is not significant during the reaction of citric acid with calcite, at least in the first few minutes where measurements were conducted.

2. In this study, the measurements of the reaction kinetics were based on the rotating disk experiments and were conducted at various high rotational speeds, where the effect of the mass transfer process is not significant. According to Taylor *et al.* (2004), the high rotational speeds reduce the boundary layer of the mass transfer process. As a result, the calcite dissolution will be limited by the acid reaction rate. In other words, the transfer rate of the acid molecules to the reacting surface is much higher than the depletion rate of the acid molecules. Therefore, in the reaction-rate limited regime,

it is valid to assume that the surface concentration of the acid is equal to its bulk concentration, $C_S = C_B$.

3. The main reactant during the reaction of citric acid with calcite is assumed to be only the citric acid molecules. As stated earlier, the pH value at the reacting surface is between 4 and 5. Therefore, it is assumed that citric acid is completely dissociated to di-hydrogen citrate and the partial dissociation of di-hydrogen citrate is insignificant.

Applying assumptions 1-3, the term (X^n . $((C_0-C_S)^{2n}/[H^+]_s^n)$) becomes zero, and Eq. (3) reduces to the following expression:

$$R = k_f \left\{ [K_{a1} * C_B]^{n/2} \right\} \qquad (4)$$

where k_f is the reaction rate constant of the forward reaction.

The temperature dependence of the reaction rate constant is usually quantified using the Arrhenius equation:

$$k_f = k_0 \exp\left\{\frac{-E_a}{R_g T}\right\} \qquad (5)$$

where k_0 is the frequency factor for the reaction, R_g is the universal gas constant (8.3144 J/mol/K), T is the temperature (K), and E_a is the apparent activation energy (J/mol). The units of the frequency factor vary depending on the reaction rate constant.

Experimental Studies

Materials

Citric acid powder was obtained from ADM Ringaskiddy Company, US Pharmacopeia (USP grade). XRD analysis showed that it contained 99 wt% anhydrous citric acid and 1 wt% mono-hydrated citric acid. Calcite marble ($CaCO_3$) with a trade name of "Acqua Bianca" was obtained from Italy. XRD analysis indicated that it contained 99 wt% calcium carbonate and trace amounts of albite, quartz, and chlorite. Calcium carbonate powder (ACS grade) was supplied by Fisher Scientific Co. Citric acid solutions were prepared using distilled water with a resistivity greater than 18MΩ/cm at room temperature. A corrosion inhibitor for organic acids was added to the citric acid at 0.1 wt% to protect surface-wetted areas. Calcium ion measurements showed that the addition of the corrosion inhibitor at 0.1 wt% did not affect the reaction rate of citric acid with calcite.

Rotating Disk Apparatus (RDA)

The RDA used in this work was RDA-100, manufactured by CoreLab Instruments Ltd. All acid-wetted surfaces were manufactured from Hastelloy B-2 or C-276 alloy for corrosion resistance.

Calcite disks with a diameter of 1.5-in and a thickness of 0.65-in were soaked in 0.1N HCl for nearly 30 min, and then thoroughly rinsed with distilled water before use. According to Fredd and Fogler (1998c), this procedure greatly improves the reproducibility of results. The calcite disk was mounted on the rotating assembly using heat-shrinkable Teflon tubing. The disk was rotated in citric acid solution with a known initial acid concentration. The experiments were run and 10 cm^3 samples of the bulk solution were collected periodically for calcium analysis. Corrections were made to account for changes in the volume of the reactants that occurred due to sample withdrawal (Taylor *et al.*, 2004). The rotating disk experiments were conducted to determine the effects of initial citric acid concentration (1, 2, 3, 5, 7.5 wt%), temperature (25, 40, 50°C), and disk rotational speed (100–1000 rpm) on calcite dissolution rate and calcium citrate precipitation. All experiments were performed at 1,000 psi. A new calcite disk was used for each experiment.

Analytical Techniques

An auto-titrator (Mettler DL70ES) was used to determine the concentration of citric acid using 0.096M NaOH up to a pH value of 8.0, using a few drops of phenolphthalein as an indicator. The concentrations of total Ca in the samples, collected from the rotating disk apparatus, were measured using inductively coupled plasma (ICP). Density measurements were made with a Paar digital density meter (DMA 35N).

To measure pH, an Orion 250A meter and Cole Parmer Ag/AgCl single junction pH electrode were used. The viscosity of citric acid solutions (Newtonian fluids) was measured using a capillary viscometer at 25, 40, and 50°C.

Results and Discussion

Fig. 1 shows the calcite dissolution rate as a function of the initial citric acid concentration at various temperatures. It also shows the calculated reaction order and different reaction constants at 25, 40, and 50°C. The average value of the reaction order was found to be **0.833**. The reaction constants were found to be 2.1E -3, 6.1E -3, and 1.6E -2 at 25, 40, and 50°C, respectively.

Based on Eq. (5), a semi-log plot of k_f vs. (1/T) will give a straight line. The slope of this line is equal to $(-E_a/(R_gT))$ and the inverse log of the y-intercept is equal to the frequency factor (k_0). From **Fig. 2**, the activation energy was calculated and found to be equal to **63.1** kJ/mol. The value of the determined activation energy is in good agreement with the reported values for calcite dissolution controlled by the surface reaction rate. A comparable value of activation energy was reported for the reaction of diluted acetic acid with calcite in phosphorite ores, 69.7 kJ/mol, (Vaimakis *et al.*, 2008). However, Fredd and Fogler (1998c) reported a lower value of activation energy, 25 kJ/mol, for the reaction of acetic acid with calcite. They found that this low activation energy value is the result of the effect of both the mass transfer and surface reaction rates on the calcite dissolution in acetic acid solutions.

Conclusions

The kinetics of the reaction of citric acid with calcite was studied using the rotating disk apparatus at 1,000 psi. Based on the results obtained, the following conclusions can be drawn:

1. A simple equation, $R = k_f \left\{ [K_{a1} * C_B]^{n/2} \right\}$, was used to model the kinetics of citric acid reaction with calcite.
2. The model parameters, n and k_f, were determined based on the calcite dissolution rates in citric acid solutions with initial concentrations of 3, 5, and 7.5 wt% at rotational speeds greater than 500 rpm
3. The average reaction order of the surface reaction of citric acid with calcite was found to be **0.833**.
4. The effect of temperature on the reaction constant was found to follow Arrhenius law.
5. The activation energy for the reaction of citric acid with calcite was found to be **63.1** kJ/mol.

Acknowledgements

The authors would like to thank the Operation and Services Division of Saudi Aramco EXPEC ARC for lending their Rotating Disk Apparatus for conducting this study. Ali Al-Aamri, Saudi Aramco EXPEC ARC and M. Al-Fahad, Saudi Aramco R&D Center, are acknowledged for conducting the rotating disk experiments and for measuring the concentration of total calcium ions, respectively.

References

AlKattan, M., Oelkers, E., Dandurand, J., Schott, J. 1998. An Experimental Study of Calcite and Limestone Dissolution Rates as a Function of pH from −1 to 3 and Temperature from 25 to 80°C. *Chemical Geology* **151**: 199–214.

AlKattan, M., Oelkers, E., Dandurand, J., Schott, J. 2002. An Experimental Study of Calcite Dissolution Rates in Acidic Conditions and 25°C in the Presence of $NaPO_3$ and $MgCl_2$. *Chemical Geology* **190**: 291-302.

Alkhaldi, M., Nasr-El-Din, H.A., Mehta, S., Al-Aamri, A. 2007. Reaction of Citric Acid with Calcite. *Chemical Engineering Science* **62** (21): 5880-5896.

Al-Mohammad, A.M., Nasr-El-Din H.A., Al-Aamri, A.M., and Al-Fuwaires, O. 2006. Reaction of Calcite with Surfactant-Based Acids. Paper SPE 102838 presented at the SPE Annual Technical Conference and Exhibition, San Antonio, TX, 24-27 September.

Al-Mutairi, S.H., Hill, A.D., Nasr-El-Din, H.A. 2008. Effect of Droplet Size and Acid Volume Fraction on the Rheological Properties and Stability of Emulsified Acid. *SPEPO* **23**(4): 484-497.

Al-Mutairi, S.H., Nasr-El-Din, H.A., Hill, A.D., Al-Aamri, A. 2008. Effect of Droplet Size on the Reaction Kinetics of Emulsified Acid with Calcite. Paper SPE 112454 presented at the SPE International Symposium and Exhibition on Formation Damage Control ,Lafayette, LA, 13–15 February.

Blauch, M.E., Cheng, A., Rispler, K. 2003. Novel Carbonate Well Production Enhancement Application for Encapsulated Acid Technology: First-Use Case History. Paper SPE 84131 presented at the SPE Annual Technical Conference and Exhibition, Denver, CO, 5–8 October.

Buijse, M., Boer, P., Breukel, B., Klos, M., Burgos, G. 2004. Organic Acids in Carbonate Acidizing. *SPEPF* **19** (3): 128-134.

Burgos, G., Birch, G., Buijse, M. 2004. Acid Fracturing with Encapsulated Citric Acid. Paper SPE 86484 presented at the SPE International Symposium and Exhibition on Formation Damage Control, Lafayette, LA, 18–20 February.

Chang, F., Nasr-El-Din, H.A., Lindvig, T., and Qiu, W. 2008. Matrix Acidizing of Carbonate Reservoir Using Organic Acids and Mixture of HCl and Organic Acids. Paper SPE 116601 presented at the SPE Annual Technical and Exhibition, Denver, Colorado, 21-24 December.

Chatelain, J.C., Silberberg, I.H., Schechter, R.S. 1976. Thermodynamic Limitations in Organic Acid/Carbonate Systems. *SPEJ* **16** (4): 189-195.

Crowe, C.W., McGowan, G.R., Baranet, S.E. 1988. Investigation of Retarded Acids Provides Better Understanding of Their Effectiveness and Potential Benefits. Paper SPE 18222 presented at the SPE Annual Technical Conference and Exhibition, Houston, TX, 2-5 October.

Fredd, C.N. 2000. Advances in Understanding and Predicting Wormhole Formation. *Reservoir Stimulation, 3rd Edition*, A-16-1. New York city: John Wiley & Sons, Inc.

Fredd, C.N. and Fogler, H.S. 1998a. Alternative Stimulation Fluids and Their Impact on Carbonate Acidizing. *SPEJ* **13** (1): 21-41.

Fredd, C.N. and Fogler, H.S. 1998b. The Influence of Chelating Agents on the Kinetics of Calcite Dissolution. *J. Colloid and Interface Science* **204** (1): 187-197.

Fredd, C.N., and Fogler, H.S. 1998c. The Kinetics of Calcite Dissolution in Acetic Acid Solutions. *Chemical Engineering Science* **53** (22): 3863-3874.

Harris, F.N. 1961. Application of Acetic Acid to Well Completion, Stimulation and Reconditioning. *JPT* **13** (7):637-639.

Huang, T., Ostensen, L., Hill, A.D. 2000. Carbonate Matrix Acidizing with Acetic Acid. Paper SPE 58715 presented at the SPE International Symposium on Formation Damage Control, Lafayette, LA, 23-24 February.

Lund, K., Fogler, H.S., McCune, C.C, Ault, J.W. 1975. Acidization II. The Dissolution of Calcite in Hydrochloric Acid. *Chemical Engineering Science* **30**: 825-835.

Nasr-El-Din H.A., Al-Mohammad, A.M., Al-Aamri, A.M., and Al-Fuwaires, O. 2008. Reaction Kinetics of Gelled Acids with Calcite. *SPEPO* 23(3): 353-361.

Nasr-El-Din H.A., Al-Mohammad, A.M., Al-Aamri, A.D., and Al-Fahad, M.A., Chang, F., 2008. Quantitative Analysis of Reaction Rate Retardation in Surfactant-Based Acids. accepted for publication, *SPEPO*.

Nasr-El-Din, H.A., Lynn J.D. and Taylor K.C. 2001. Lab Testing and Field Application of a Large Scale Acetic Acid Based Treatment in a Newly Developed Carbonate Reservoir. Paper SPE 65036 presented at the SPE Oilfield Chemistry, Houston, TX, 13-16 February.

Plummer, L.N., Wigley, T.M., and Parkhurst, D.L. 1978. The Kinetics of Calcite Dissolution in CO_2 Water Systems at 5°C to 60°C and 0.0 to 1.0 Atm. CO_2. *J. American Science* **278**: 179-216.

Robert, J.A. and Crowe, C.W. 2000. *Reservoir Stimulation*, 3rd Edition, 17-19. John Wiley & Sons Inc.: 17.

Taylor, K.C., Al-Ghamdi, A., Nasr-El-Din, H.A. 2004. Effect of Additives on the Acid Dissolution Rates of Calcium and Magnesium Carbonates. *SPEJ* **19**: 122–127.

Taylor, K.C., Nasr-El-Din, H.A., Mehta, S. 2006. Anomalous Acid Reaction Rates in Carbonate Reservoir Rocks. *SPEJ* **11**(4):488-496.

Vaimakis, T., Economou, E., Trapalis, C. 2008. Calorimetric Study of Dissolution Kinetics of Phosphorite in Diluted Acetic Acid. *J. Thermal Analysis and Calorimetry* **92** (3): 783-789.

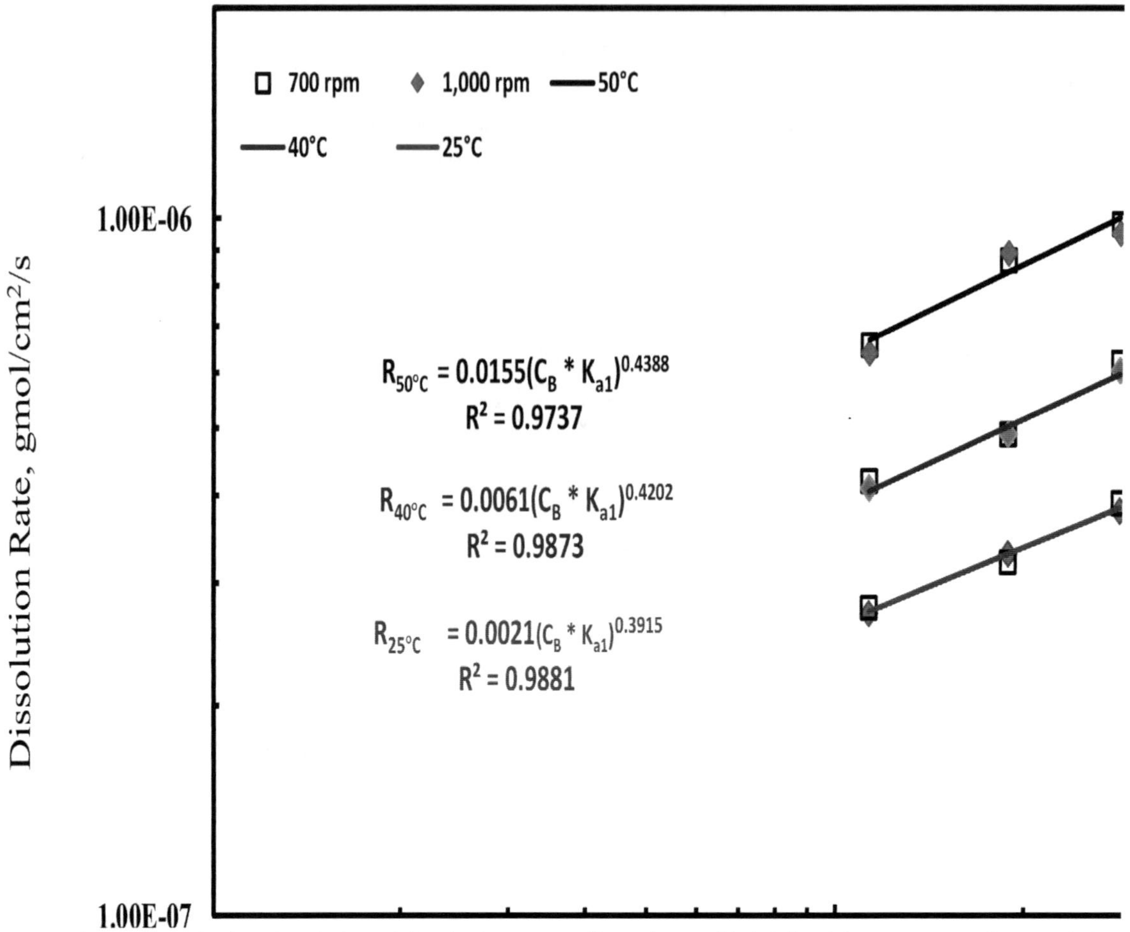

Fig. 1: Calcite dissolution in citric acid solutions as a function of initial acid concentration at 25, 40, 50°C and 1,000 psi.

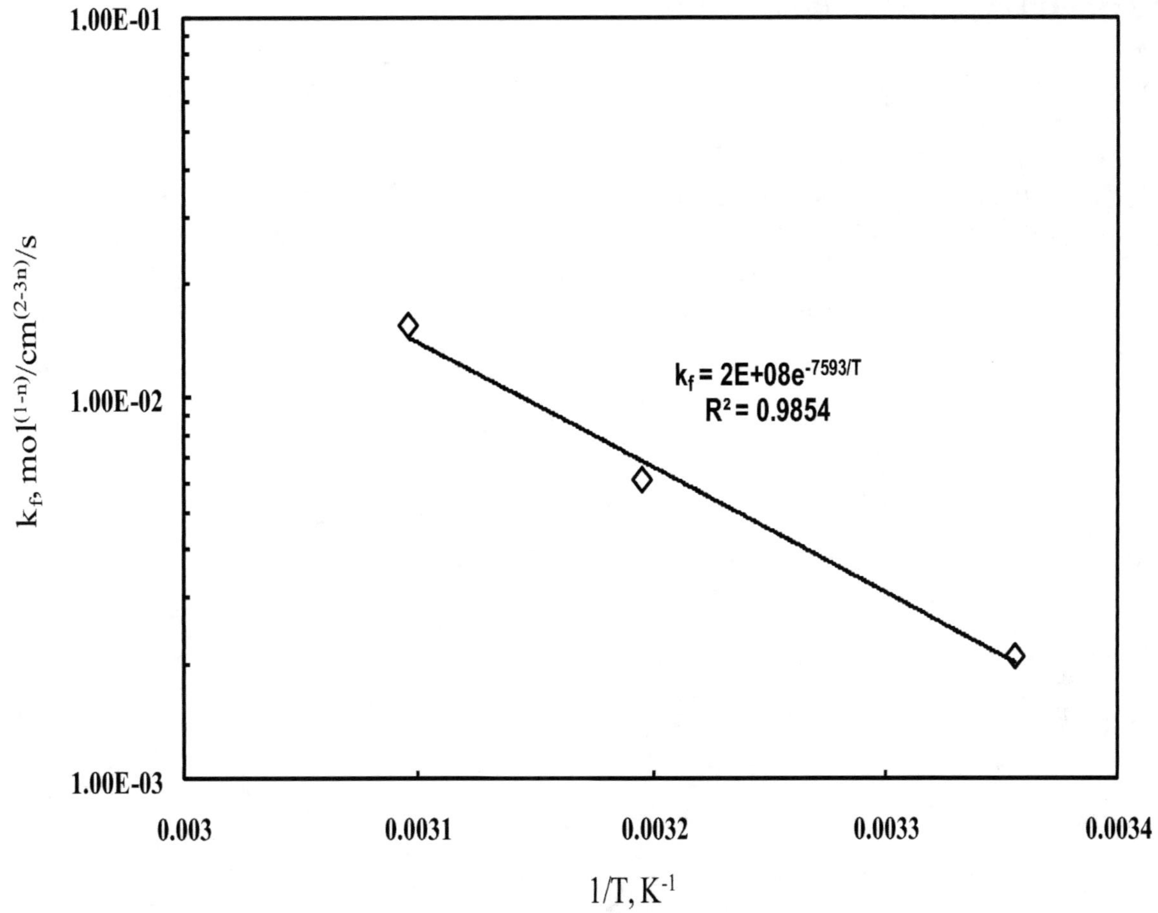

Fig. 2: Reaction rate constant as a function of reaction temperature.

53f Analysis of the Darcy Continuity Equation

Jonathan Worstell, Hoder, Shell Global Solutions US, 3333 Highway 6 South, Room CP-109, Houston, TX 77082

Enhanced Oil Recovery involves flow through porous media. As chemical engineers, we design processes for liquid flow and gas flow through pipes and ducts. We readily understand such flows; however, we do not generally encounter flow through porous media when designing a chemical process. This talk discusses the conservation equations and boundary conditions for flow through porous media, as related to Enhanced Oil Recovery and chemical engineering.

68a Review of State-of-the-Art in Biomass Gasification Technologies (Webcast Not Available)

Mahesh V. Iyer, Chemical Process Technology, Shell Global Solutions (US) Inc., Westhollow Technology Center, 3333 Highway 6 South, Houston, TX 77082 and Joseph B. Powell, Process Development GSICP/1, Shell Global Solutions US Inc., Westhollow Technology Center, P. O. Box 1380, Houston, TX 77251-1380.

A review of current biomass gasification processes with process development issues will be presented. The focus will be on various thermochemical routes for applications ranging from power production, hydrogen manufacture and syngas conversion for liquid fuel synthesis

68b Investigation of Effects of Coal and Biomass Contaminants on the Performance of Water-Gas-Shift and Fischer-Tropsch Catalysts

Gökhan Alptekin, Ambalavanan Jayaraman, Margarita Dubovik, Matthew Schaefer, Robert Amalfitano, and Lauren Brickner. TDA Research Inc., 12345 W. 52nd Ave, Wheat Ridge, CO 80033

TDA Research, Inc with Department of Energy (DOE) funding is carrying out an investigation of the effects of coal and biomass contaminants on the performance (activity and selectivity) and life of the Water Gas Shift (WGS) and Fischer Tropsch (F-T) catalysts. The poisoning effects of heteroatoms (e.g., sulfur, nitrogen and oxygen) as well as halides and trace metals, are well-documented for these catalysts. In the co-gasification of coal and biomass, it is anticipated that ash, sulfur species, trace toxic metals, halides, and nitrogen species will be lower than in the synthesis gas generated by coal alone. Nevertheless, the impact of these contaminants must be re-evaluated at the reduced concentrations predicted for the intended application. In addition, the biomass feedstock will introduce a different spectrum of contaminants, including the vapors of alkali metals (such as sodium and potassium compounds) and their respective salts (e.g., KCl and NaCl), ammonia and phosphorous that could render the WGS and FT catalysts inactive.

We have first used thermodynamic modeling tools to identify contaminants that are most likely to react with the active phase and promoters used in the WGS and F-T catalysts. We then carried out screening experiments utilizing generic and commercial catalysts to measure changes in activity and selectivity due to exposure to contaminant species under representative conditions. In this paper, we will present the preliminary results from the catalyst development and contaminant evaluations done at TDA.

Fuels from Biomass Via High Temperature Water

Joseph B. Powell, Ph.D.
Chief Scientist – Chemical Engineering
Shell Global Solutions U.S. Inc.
Westhollow Technology Center, Houston, TX

Summary

To date, water treatment of biomass has been examined across a broad spectrum of parameter space, from hot liquid water (HLW) to near critical or supercritical (SCW), in an effort to either directly synthesize liquid or gaseous chemicals and biofuels, or to pretreat biomass to allow further processing.

At temperatures between 150 and 200 °C in liquid hot water (LHW), hemicellulose is extracted as xylan and hydrolyzed to produce chemical coproducts xylitol and furfural (Ando 2000). Some lignin may be removed (depending on biomass type), and the resulting modified cellulosic solids may be subjected to pulping, or enzymatic hydrolysis for ethanol fermentation (Liu 2005). Further increase in temperature leads to degradation of xylans to furfural (Laser 2002), with extraction of glucans or celluloses, and additional lignin. Ultimately, 85% of the biomass may be solubilized as temperature is increased to 270 °C (Ando 2000). Solvolysis reactions in this regime include isomerization, dehydration, fragmentation, and condensation reactions (Matsumura 2005) to produce a complex mixture of alkyl phenols, furans, carbohydrate dimers, and acids (Garcia-Perez et al. 2007). Many of these compounds are inhibitors for enzymatic hydrolysis of cellulose, or ethanol fermentation.

Slow pyrolysis at intermediate temperatures (200 – 350 °C) leads to formation of "tars" and irreversible formation of char, giving low yields of fuels from biomass (Matsumura 2005). Rapid hydrothermal treatment above 350 °C (near or sub-critical water) can generate higher yields of gas-phase products from biomass. A catalyst is typically needed for gasification in the region 350 – 550 °C (Elliott 2008). Supercritical water gasification can proceed non-catalytically above 500 C, yielding gas-phase products with negligible tar or char, as modeled via cellulose gasification (Guo et al. 2007):

(1) $C_6H_{10}O_5 + 7\ H_2O \rightarrow 6\ CO_2 + 12\ H_2$

H_2 becomes the principal product above 600 C due to favorable equilibrium at high temperature relative to the reverse water gas shift reaction:(Savage 1999, 2009)

(2) $CO_2 + H_2 \leftrightarrow CO + H_2O$

Beneficial use of water in biomass processing is thus limited to low temperature (< 200 °C) removal of hemicellulose for production of chemicals, leaving accessible cellulose for enzymatic hydrolysis and ethanol fermentation (Hendriks 2009) or pulping, and high temperature, fast gasification for production of fuel gases. The latter entails separate catalytic 350 – 600 °C and noncatalytic supercritical (> 600 °C) regions, which differ in design requirements and operational challenges. The high temperature regions make use of the unique properties of water at elevated temperatures, where dielectric constant diminishes in the vicinity of the 374 °C critical temperature to provide solvation similar to organic solvents at room temperature, and the ionization constant increases at the critical point to increase specific acid / base catalysis by three orders of magnitude (subcritical water), before rapidly decreasing again to provide a supercritical temperature regime largely free of ionic reaction (Savage, 2009).

A simple screening apparatus was assembled to test near or supercritical water thermolysis of biomass (Figure 1). The unit entails high alloy (Inconel) tubing ¼-inch diameter, with welded bottom, rated to 7000 psi at 600 °C, and fitted with rupture disk. The unit is loaded with biomass (walnut shells or glycerol) and a desired amount of water, leaving a partial headspace for gas collection. The headspace may be pre-filled with N_2 to achieve faster transition to supercritical water pressure upon heat-up. The "Bioroaster" is immersed half-way into a fluidized sand bath. Table 1 shows results for preliminary results where the bath is programmed from room temperature, realizing 60% of set point temperature in 30 minutes. Alternate fast immersion runs for preheated bath (#27 & 28) achieved supercritical water conditions within 20 seconds. Resulting gas generation and selectivities indicate gasification of only 4 – 17% of the available biomass feed relative to theoretical gas generation of equation 1, with substantial residual oil and char formation. Similar low gaseous product yields were reported by Garcia-Perez (2007) for high temperature gasification at essentially atmospheric pressure. Fast immersion for preliminary tests did not generate substantially more gaseous products in preliminary testing. Operation at supercritical temperature but sub-critical pressure yields a regime where water exhibits very low ionization constant and specific acid/basis catalysis is suppressed (Yagasaki 2005), while intermediate transient regimes entail liquid water acting at enhanced acidity at substantial temperatures. Future improvements will entail use of glass insert to quantify residues as a function of gasification conditions.

Commercial operation with water at or above critical temperature and pressure is challenged by corrosion (Antal 2000), and by requirements for strength of materials at high temperatures and pressures. Use of very high temperatures requires effective heat recovery in a continuous flow system (Matsumura 2005), in order to have a positive energy balance. A further commercial challenge is the potential requirement for rapid heat transfer to a slurry of solid biomass, to avoid a transition region where nonrevertible char and tar formation can occur. Further delineation of these requirements (temperature, pressure, and residence time in transition regions) is needed to assess commercial viability.

References

Ando, H., Sakaki, T., Kokusho, T., Shibata, M., Uemura, Y., and Hatate, Y. 2000. Decomposition Behavior of Plant Biomass in Hot-Compressed Water. *Industrial & Engineering Chemistry Research* 39 (10):3688-3693.

Antal, M.J. and, S. G. Allen, D. Schulman, X. Xu, R. J. Divilio. 2000. Biomass Gasification in Supercritical Water. *Ind. Eng. Chem. Res., 2000, 39 (11), 4040-4053.*

Elliott, D. C. 2008. Catalytic hydrothermal gasification of biomass. *Biofuels, Bioproducts and Biorefining* 2 (3):254-265.

Garcia-Perez, Manuel, Thomas T. Adams, John W. Goodrum, Daniel P. Geller, and K. C. Das. 2007. Production and Fuel Properties of Pine Chip Bio-oil/Biodiesel Blends. *Energy & Fuels* 21 (4):2363-2372.

Guo, L. J., Y. J. Lu, X. M. Zhang, C. M. Ji, Y. Guan, and A. X. Pei. 2007. Hydrogen production by biomass gasification in supercritical water: A systematic experimental and analytical study. *Recent Advances in Catalytic Production of Hydrogen from Renewable Sources* 129 (3-4):275.

Hendriks, A.T.W.M, and Zeeman, G. 2009. Pretreatments to enhance the digestibility of lignocellulosic biomass. *Bioresource Technology* 100:10-18.

Laser, Mark, and D. Schulman, S. G. Allen, J. Lichwa, M. J. Antal, L. R. Lynd. 2002. A comparison of hot liquid water and steam pretreatments of sugar cane bagasse for bioconversion to ethanol. *Bioresource Technology* 81:33-44.

Liu, C, and C. E. Wyman. 2005. Partial flow of compressed-hot water through corn stover to enhance hemicellulose sugar recovery and enzymatic digestibility of cellulose. *Bioresource Technology* 96:1978-1985.

Matsumura, Y., and T. Minowa, B. Potic, S. R. A. Kersten, W. Prins, I. P. M. van Swaaij, B. van de Beld, D. C. Elliott, G. G. Neuenschwander, A. Kruse, M. J. Antal Jr. 2005. Biomass gasification in near- and super-critical water: Status and prospects. *Biomass and Bioenergy* 29 (4):269-292.

Savage, Phillip E. 1999. Organic Chemical Reactions in Supercritical Water. *Chemical Reviews* 99 (2 %R doi:10.1021/cr9700989):603-622.

———. 2009. A perspective on catalysis in sub- and supercritical water. *20th Year Anniversary Issue of the Journal of Supercritical Fluids* 47 (3):407.

Yagasaki, T, and K. Iwahashi, S. Saito, I. Ohmine. 2005. A theoretical study on anomalous temperature dependence of pK_w of water. *J. Chem. Phys. 122 (2005).* 122:144504.

Figure 1: "Bioroaster" for rapid screening of Near & Supercritical Water Biomass Hydrolysis

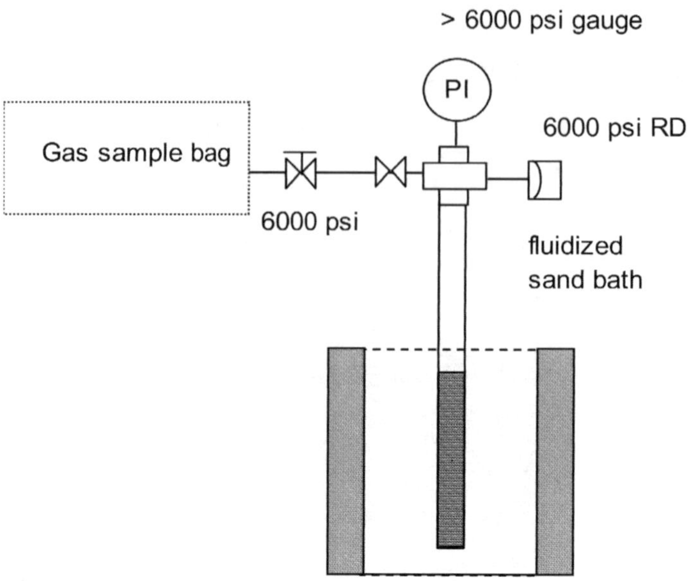

Table 1: Preliminary Tests of Near and Supercritical Water Thermolysis

6:1 to 8:1 water/ biomass (by weight)

#	Feed	ramp (hrs)	final T (C)	P (psi)	total Hours	*Conv % biomass	gas product %H2	gas product %CH4	gas product %CO2	gas product %CO
26639-8	Walnut	2.0	570	1507	4.90	10.2%	6.4%	35.5%	44.8%	13.3%
26639-9	Walnut	2.0	584	4815	3.10	4.8%	18.3%	39.3%	35.6%	6.7%
26639-11	Walnut	2.0	523	5238	1.50	7.5%	15.3%	33.0%	43.9%	7.9%
26639-13B	Glycerol	2.0	540	5906	4.67	15.3%	5.7%	19.6%	49.0%	25.7%
26639-16	Walnut	2.0	550	4900	3.50	13.2%	14.8%	35.2%	45.1%	4.9%
26639-27	Glycerol	0.03	550	5000	0.03	17.9%	n.a	n.a	n.a	n.a
26639-28	Walnut	0.03	550	5000	0.03	9.0%	n.a	n.a	n.a	n.a

*% of maximum theoretical gas product basis cellulose hydrolysis equation #1

Slow temperature ramp with transition to 60% of set point in 0.5 hours (runs 8 – 16). Fast immersion (runs 27,28) gave set point in 2 minutes and exceeded SC water conditions within 20 seconds.

68d Kinetic Modeling of Oxidative-Short-Term Lime Pretreatment of Biomass

Rocio Sierra, Ingenieria Quimica, Universidad de los Andes, Edificio Mario Laserna, Cra. 1 Este 19 A 40, Bogota, Colombia and Mark Holtzapple, Chemical Engineering, Texas A&M University, Chemical Engineering, 3122 TAMU, College Station, TX 77843.

Oxidative lime pretreatment has shown effective to increase digestiblity of lignocellulosic resources. Kinetic models for oxygen delignification have been developed to gain insight into the pretreatment method. These models can be used in the design of comercial-scale equipment and process control and optimization.

68e Conversion of Lignocellulose and Other Waste Biomass into Ketones and Hydrocarbon Biofuels: The MixAlco Process, Semi-Works Plant Results

Cesar B. Granda[1], Mark T. Holtzapple[2], Gary W. Luce[3], and Ray H. Griggs[3]. (1) Chemical Engineering Dept., Texas A&M University, M.S. 3122, Jack E. Brown Bldg., College Station, TX 77843, (2) Chemical Engineering Depart., Texas A&M University, M.S. 3122, Jack E. Brown Bldg., College Station, TX 77843, (3) Terrabon, L.L.C., 20333 State Highway 249, Suite 200, Houston, TX 77070

The MixAlco process converts any anaerobically biodegradable material (e.g., proteins, cellulose, hemicellulose, fats, pectin) into a wide array of chemicals and fuels. Such conversion occurs by anaerobic fermentation of the biomass into mixed acids by a mixed culture of naturally occurring microorganisms followed by the conversion of the mixed acids into the desired chemicals or fuels using conventional chemistry. Terrabon, LLC built a semi-works plant in Bryan, TX to confirm the scalability of this process. The semi-works plant processes the equivalent to 5 to 10 ton/day of biomass and is able to generate enough fermentation products to produce 300 gal/day of bio-gasoline. Such conversion to bio-gasoline takes place in a separate pilot plant, where the fermentation product is clarified, dewatered, thermally converted into ketones, and finally catalytically oligomerized into hydrocarbons. The semi-works plant monitors different parameters in the fermentation, such as conversion, selectivity towards acids and acid yield and biogas production. In the downstream chemical processing into hydrocarbons, many different parameters will be monitored, such as yields and selectivity in the thermal conversion to ketones and oligomerization to hydrocarbons, as well as the quality (e.g., octane number, RVP and emissions) of the final product. Results from these studies will be presented.

Web Page: www.terrabon.com/

68f Recovery of Chemicals Produced by Acidogenic Fermentation

Emilie Fockedey, Imap, Universite catholique de Louvain (UCL), Reaumur, Place Sainte Barbe, 2, Louvain-la-Neuve, B - 1348, Belgium, Patrick Gerin, Gebi, Universite catholique de Louvain (UCL), Reaumur, Place Sainte Barbe, 2, Louvain-la-Neuve, B - 1348, Belgium, and **Juray De Wilde**, Materials and Process Engineering Dept. (IMAP), Universite catholique de Louvain (UCL), Reaumur, Place Sainte Barbe, 2, Louvain-la-Neuve, B - 1348, Belgium.

The use of alternative and renewable resources, such as biomass, for the production of energy and chemicals is motivated by decreasing oil reserves and environmental concerns. Acidogenic fermentation of biomass for the production of ethanol and acetic, propionic, butyric and lactic acids offers major advantages in that it allows: (i) non germ free conditions; (ii) using very diversified, less pretreated and cheap biomass; (iii) producing molecules and corresponding esters consumed on a large scale by the industries.

The product recovery, separation and concentration steps downstream of the biomass transformation remain a major challenge for the application of acidogenic fermentation which produces very dilute (concentrations of only a few percent) and complex aqueous solutions of acids and ethanol. A number of separation processes can be potentially used, but a process which is technically, energetically and environmentally promising has not been developed yet. In this work, a novel extraction - re-extraction process is presented which is based on the transfer of the acids and ethanol to a glycerol phase via an intermediate solvent phase. In the first, extraction step, tri-n-octylamine based solvents are chosen for their preferential extraction of acids. In the second, re-extraction step, glycerol is used to extract the acids from the intermediate solvent. Glycerol being a by-product of the growing biodiesel production, its use opens perspectives for the transformation of glycerol into short chain esters, directly using the acids recovered.

The mass based distribution coefficients were experimentally measured. Both model solutions and real fermentation broths were used. Solutions were stirred to reach equilibrium and then centrifuged. The concentrations of butyric acid in the aqueous phase before and after extraction were determined by gas chromatography. The influence of the pH, the temperature and the composition of the aqueous phase on the recovery of the acids was studied. Furthermore, the influence of the composition of the organic phase, i.e. different modifiers and diluents, was investigated.

Acknowledgements

The authors would like to acknowledge the Belgian Fonds pour la formation à la Recherche dans l'Industrie et dans l'Agriculture (F.R.I.A.) for the financial support.

References

1. C. O'Driscoll (2007) Biofuels, Bioprod., Bioref., 1, 6–7

SOLVENT EVALUATION FOR ECOGASOLINE PRODUCTION VIA TWO STEP PROCESS

Alberto J. Baumeister, Sebastiano Giardinella and Katiuska C. Márquez, Process Engineering, INELECTRA PANAMA, S.A., Panama, Panama

Introduction and Background

The use of ethanol as an alternate fuel is seeing widespread acceptance due to growing environmental concerns, fluctuating oil prices and political interest of certain governments towards gaining energetic independence, among other reasons. Around 40 billion litres (11 billion US gallons) of fuel ethanol were produced in the world in the year 2006, out of which about 90% was produced in the United States and Brazil [1]. The increase in ethanol demand in the last decade is evident: in the year 2006 the demand for fuel ethanol in the US reached 20 billion litres (5.4 billion US gallons), more than twice the demand for fuel ethanol in the year 2002, of 7.9 billion litres (2.1 billion US gallons) [2].

Fuel ethanol is blended with petroleum products to produce the commercially available ethanol fuel blends. Two ethanol-hydrocarbon blends can be purchased in fuel dispensers across the U.S.: E10 (available in many common gasoline pump stations), which is 10% ethanol and is compatible with most automotive motors, and E85 (Ed75-Ed85, according to norm ASTM D-5798), which is 85% ethanol and compatible with dual-fuel motors.

Several studies, however, have pointed-out to disadvantages in the primary processes for ethanol production in the US. These disadvantages include: increase in basic grains and fertilizer prices because of competition between the fuel and food markets, debatable energetic efficiency (it has been argued that US corn ethanol requires more energy than it produces, when considering the entire production costs of farming, fertilizer, seeds, pesticides, transport, distillation and others [3]; in contrast, Brazilian sugarcane ethanol has been reported as both sustainable and efficient [4]), among others. In order to find solution to the aforementioned problems, several alternatives are being developed which include: production of ethanol from non-eatable materials such as biomass or cellulose, or from other sources like algae. In parallel, another trend of research that has taken resent interest is the development of alternative processes to purify ethanol.

The primary methods to produce ethanol require the separation of ethanol from water, which is traditionally done by distillation, an energy-intensive process. The maximum ethanol purity that distillation can produce is given by the ethanol-water azeotrope, and additional processing is needed to produce fuel-grade ethanol; these processes include: vacuum distillation, azeotropic or extractive distillation and mole sieve separation.

Several authors have proposed solvent extraction as a possible means of separating ethanol from water, and a wide variety of laboratory studies have been developed in order to obtain triangular equilibrium diagrams, distribution coefficients and separation factors in order to limit the number of potential solvents. The use of gasoline and other petroleum derivatives has been suggested as a possible means to produce gasohol (a 10% ethanol mixture) in a once-through operation [5] [6]. Fatty acids present in vegetable oils, such as olive, coconut oil, among others, have also been extensively studied as possible solvents for ethanol extraction: castor oil has been reported, among the commercial vegetable oils, as an efficient extraction agent [7] [8], whereas nonanoic acid, as a pure compound, has also been considered [9]. Significant energy savings have been reported for a simulated extraction and flash separation process using nonanoic acid over conventional distillation to produce 69.5% wt ethanol [9].

In this study, the potential of four different solvents: n-octane, diesel, oleic acid and nonanoic acid as separation agents for dilute mixtures of ethanol in water with the final purpose of producing ethanol fuel mixtures (i.e. E10, E85) was tested with the use of commercial simulation software. The objectives of this study included: a) testing the reproducibility of experimental liquid-liquid equilibrium data with commercial simulation software using available thermodynamic methods, comparing the predicted to the reference experimental values and selecting the most appropriate thermodynamic method(s) for the following ternary systems: ethanol-water-octane, ethanol-water-diesel, ethanol-water-nonanoic acid and ethanol-water-oleic acid; b) simulating the separation of ethanol from water combining an extraction stage using each of the aforementioned solvents with a distillation column; and c) estimating the operational parameters, such as solvent-to-feed ratio, number of stages and energy consumption, and comparing them to existing conventional processes to obtain ethanol fuel mixtures.

Process Simulation and Calculation Procedure

Ternary ethanol – water – solvent equilibria were simulated with PRO/II ® simulation software. Different built-in liquid activity based LLE thermodynamic packages were selected and tested against experimental data for the following systems: ethanol-water-octane, ethanol-water-diesel, ethanol-water-nonanoic acid and ethanol-water-oleic acid. The references for each set of experimental data are given on **Table 1**. The single components ethanol, water, n-octane and n-nonanoic acid were selected from the software databank; diesel was simulated as a pure component (n-dodecane) for simplification; oleic acid was defined as a mixture of fatty acids present in the software databank with compositions shown on **Table 2**.

Ethanol-water-solvent streams were created and fed to a separator unit with two liquid outlets at the conditions reported by the references. Where experimental binodal curves and tie line data were available, the simulations were carried out over a wide range of compositions in order to obtain the triangular equilibrium diagrams, which were then superimposed to the reference diagrams for comparison. Where experimental separation factors and distribution coefficients were reported, the simulations were performed by setting equal initial mixture compositions as those set by the consulted references, and both values were later computed from the simulated exit stream compositions. The thermodynamic package that most accurately reproduced the experimental data was chosen for each system and used as base for further simulations. When either of the default thermodynamic packages fitted the experimental data with accuracy, it was selected; if not, then a predictive method (i.e. UNIFAC or either of its modifications) was used to fill the binary interaction parameters; if this too proved unsatisfactory, then the available binary interaction parameter data (i.e. mutual solubilities) for the selected activity model (i.e. UNIQUAC) was input. As a last resort, if neither approach produced adequate results, then the distribution coefficients from literature were manually set on the simulator.

Table 1. Experimental Data References

System	Distribution coefficient / Separation Factor	Temperature, °C	Source	Remarks
Ethanol-water-octane	0.073 / 40.56	25.0 ± 0.1	[6]	At 44.1 wt% ethanol initial mixture concentration
Ethanol-water-diesel	0.0172 / 86.0	25.0 ± 0.1	[6]	At 16.5 wt% ethanol initial mixture concentration
Ethanol-water- nonanoic acid	0.35 / -	25 ± 1	[9]	At 7.81 g/dL of ethanol initial mixture concentration. 1.5 g/dL of water in solvent phase were reported
Ethanol-water-oleic acid	0.0458 / 21.6	33	[8]	At 5% wt ethanol initial mixture concentration
	0.069 / -	25 ± 1	[9]	At 7.81 g/dL of ethanol initial mixture concentration. Negligible water in solvent phase was reported

Table 2. Oleic Acid Composition

Component name	Mass %
Myristic acid	0.65
Palmitic acid	11.50
Stearic acid	2.00
Oleic acid	75.00
Linoleic acid	9.50
Linolenic acid	1.35

Following validation of the thermodynamic models, process simulations were developed combining a countercurrent solvent extraction column followed by a distillation column or viceversa. In order to establish a common base for comparison, equal feed composition, ethanol recovery and operating conditions were set, which are given on **Table 3**. The process flow diagrams are shown on **Figure 1**. The target product qualities were defined from commercial fuel ethanol mixtures. A target ethanol recovery ratio of 95% was chosen for all simulations. A conventional distillation operation was simulated in order to determine a heat consumption base value. The distillation operation was simulated to the maximum attainable purity for that solvent in order to compare energy requirements.

Table 3. Process Simulation Base Parameters

Parameter	Value
Feed flowrate, kg/h	10,000
Feed weight percentage, ethanol	8%
Feed temperature, °C	25
Distillation tower top pressure, kPa	101.325
Maximum water content in E10, Ed75-Ed85, mass %	1.0
Minimum ethanol content in E10, volume %	10
Minimum ethanol content in Ed75-Ed85, volume %	70*
Target ethanol recovery	95%

*Minimum volume % for Class 3 Fuel Ethanol

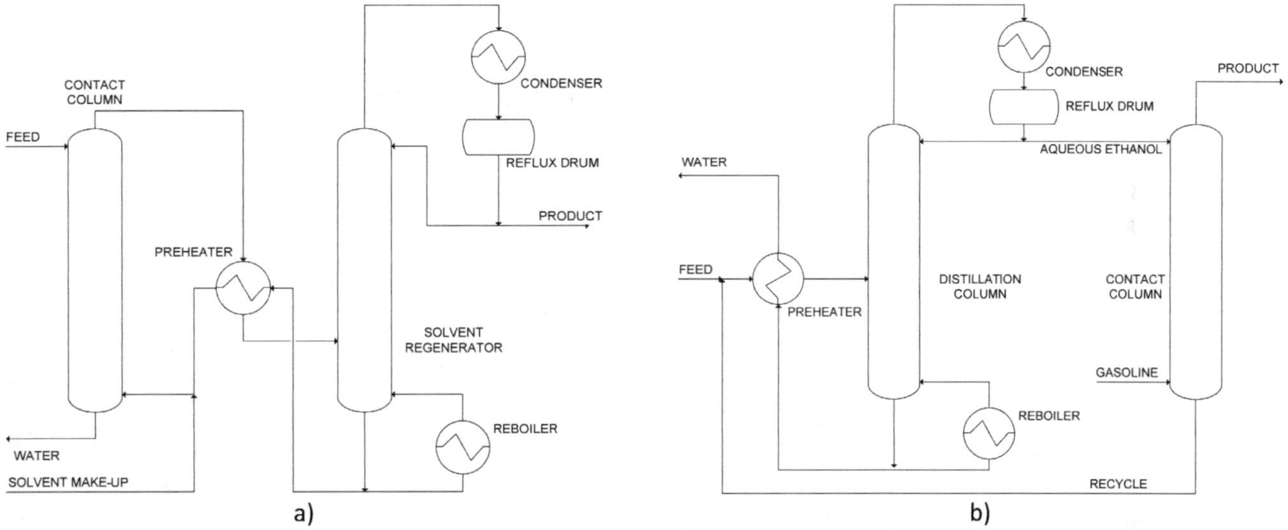

Figure 1. Process Flow Diagrams
a: extraction followed by distillation
b: distillation followed by extraction

Results and Discussion

Thermodynamic Method Selection

The UNIFAC thermodynamic model allows predicting the ternary equilibrium data for the ethanol-water-octane system at 298 K. The binodal curves and tie lines superimpose the data reported by [6] as can be seen on **Figure 2**. The distribution coefficients and separation factors reported by [6] at different feed compositions were reproduced with relative differences below 5% and 2%, respectively. The simulated distribution coefficient for ethanol in octane, based on an 8% weight ethanol in water solution at 25 °C was 0.11 with a separation factor > 20.

The overall shape of the ethanol-water-diesel system binodal curve and tie lines was reproduced with the UNIFAC model, simplifying the solvent as n-dodecane, as can be seen on **Figure 2**; however, the predicted ethanol composition in the solvent phase was higher than that reported by

[6]. [6] reported ethanol concentrations below 1 % wt in the solvent phase, whereas the simulator predicted ethanol concentrations slightly above 1 % wt. The predicted distribution coefficient for ethanol in "diesel" (simplified as n-dodecane), based on an 8% weight ethanol in water solution at 25 °C was 0.09 with a separation factor >20.

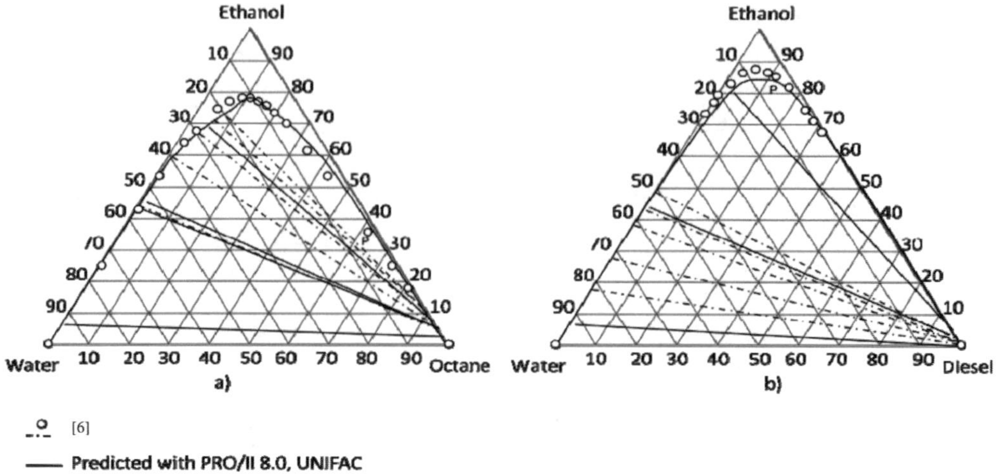

Figure 2. Triangular equilibrium diagrams for ethanol-water-solvent systems at 25.0 ± 0.1 °C
a: ethanol-water-octane
b: ethanol-water-diesel

On the ethanol-water-nonanoic system, none of the available thermodynamic packages was able to reproduce, alone, the distribution coefficient and water absorption measured by [9] of 0.35 ± 6.4% and 1.5 g/dL ± 40% at 25 ± 1 °C, respectively. Due to the lack of experimental data in the consulted reference it was not possible to regress the binary interaction parameters for UNIQUAC or NRTL models; however, it was found that by selecting UNIQUAC as the thermodynamic model, filling the missing parameters with UNIFAC and manually adjusting the mutual solubilities of the water / nonanoic acid pair, the reference distribution coefficient of 0.35 and water absorption value of 1.5 g/dL were predicted by the simulator with the solvent-to-feed volume ratios used by [9] on their experiments.

Similar to the previous system, the distribution coefficient for the ethanol-water-oleic acid system could only be reproduced by manually adjusting the mutual solubilities. In this case, not only the mutual solubility of the water / solvent pair was set, but also that of ethanol / solvent, since the simulator predicted a larger distribution coefficient than any of the consulted references. The selected thermodynamic model was also UNIQUAC with missing parameters filled with UNIFAC. In this case, the consulted references reported differing distribution coefficients for this solvent, which can be explained in the complexity of the vegetable oil and in varying compositions of the different available references. [9] reported a distribution coefficient of 0.069 ± 40% at 25 ± 1 °C with negligible water absorption for commercial grade monounsaturated oleic acid from Emery Industries Ltd, whereas [8] reported a distribution coefficient of 0.0458 and a separation factor of 21.6 for edible olive oil from Welch, Holme & Clark. Two different models were adjusted to match the data of both references.

Process Simulations Results

The simulation results for the liquid-liquid extraction processes under the scheme shown on **Figure 1a** are summarized on **Table 4**. The number of stages and feed tray location for the distillation column were determined in order to minimize energy consumption. The simulation results for the liquid-liquid extraction process under the scheme shown on **Figure 1b** are given on **Table 5**. For this scheme only gasoline (simplified as octane) was used, since it was the only solvent able to produce a commercial product (E10) at this configuration in a once-through process. The energy consumption for some available ethanol / water separation technologies is shown on **Table 6**. In this case the energy consumption for the first process stage (distillation) was simulated, and the energy consumption for the second stage (dehydration) was taken from [10].

As can be seen on **Table 4**, none of the solvents were able to produce, under the scheme shown on **Figure 1a**, either E85 or fuel grade anhydrous ethanol, as the weight percentage of water exceeded the allowable limit. An additional finishing stage would be required in order to remove the excess water. Oleic acid was the least favorable solvent, as it required a larger solvent-to-feed mass ratio, and was also more difficult to remove, since the ethanol concentration in the solvent was very low. Nonanoic acid, on the other hand, required the lowest solvent-to-feed ratio and even though the ethanol product purity was limited to 70% wt., it required 27% less energy than distillation alone to achieve it. The hydrocarbon solvents had the advantage of absorbing little water and not requiring complete removal in order to produce E85, but had high solvent-to-feed ratios.

On the other hand the process scheme shown on **Figure 1b**, which is similar to the process depicted on [5], was able to produce E10 within specifications requiring less energy than the other ethanol dehydration processes shown on **Table 6**, and consumes 16% less energy than distillation followed by molecular sieves. The main advantage of this process is that it does not need distilling the ethanol/water mixture up to the azeotrope, and no further purifying stages are involved to meet the E10 specifications.

Table 4. Liquid-Liquid Extraction Process Simulation Results (Scheme 1a)

Solvent	Extraction Column		Solvent Recovery Column			Product Composition			Reboiler
	Solvent-to-Feed Mass Ratio	Number of Stages	Number of Stages	Feed Stage	Reflux Ratio	Ethanol wt%	Water wt%	Solvent wt%	Duty M J/kg ethanol
Octane	8.7	11	14	7	5.0	68.9	1.9	29.2	7.1
Diesel	10.8	11	10	5	3.8	70.9	2.0	27.1	4.5
Nonanoic Acid	3.1	10	10	5	3.2	70.0	30.0	0.0	2.8
Oleic Acid*	15.8	10	14	10	7.2	96.4	3.6	0.0	6.7

* Assuming a distribution coefficient of 0.069

Table 5. Liquid-Liquid Extraction Process Simulation Results (Scheme 1b)

Solvent Recovery Column			Extraction Column		Product Composition .			Reboiler
Number of Stages	Feed Stage	Reflux Ratio	Solvent-to-Feed Mass Ratio	Number of Stages	Ethanol wt%	Water wt%	Solvent wt%	Duty M J/kg ethanol
14	9	2.9	5.5	10	11.2	0.2	88.6	5.7

Table 6. Energy Consumption for Conventional Ethanol / Water Separation Processes

Process Stage 1 / 2	Feed Ethanol wt% Stage 1 / 2	Product Ethanol wt% Stage 1 / 2	M J/kg ethanol Stage 1 + 2* = Total
Distillation / -	8 / -	70 / -	3.8
Distillation / Azeotropic Distillation with Benzene	8 / 93.5 (85 mole%)	93.5 (85 mole%) / Anhydrous	4.8 + 4.9* = 9.7
Distillation / Extractive Distillation with Ethylene-Glycol	8 / 93.5 (85 mole%)	93.5 (85 mole%) / Anhydrous	4.8 + 1.8* = 6.6
Distillation / Extractive Distillation with Gasoline	8 / 93.5 (85 mole%)	93.5 (85 mole%) / Anhydrous	4.8 + 3.2* = 8.0
Distillation / Adsorption with Molecular Sieves	8 / 95.5 (88 mole%)	95.5 (88 mole%) / Anhydrous	5.4 + 1.4* = 6.8

* The energy consumptions for azeotropic distillation with benzene, extractive distillation with ethylene-glycol, extractive distillation with gasoline and adsorption with molecular sieves were taken from [10].

Conclusions

1. The UNIFAC thermodynamic model reproduced the reference experimental data for the ethanol-water-octane and ethanol-water-diesel systems.
2. The UNIQUAC thermodynamic model, filled with UNIFAC and slightly adjusted to match the experimental data was able to reproduce the ethanol-water-nonanoic acid reference experimental data. This model was also used for the ethanol-water-oleic acid system, but in this case due to the nature of the oleic acid mixture, the results are not comparable because of differing compositions.
3. Processes for obtaining ethanol fuel mixtures by combining an extraction stage followed by a distillation stage were simulated whereby the modeled schemes were unable to produce E85 fuel under specifications with octane, diesel, nonanoic acid or oleic acid.

4. A distillation column followed by extraction with gasoline was simulated and was able to produce E10 within specifications and with lower energy consumption than the compared available technologies.

References

1. The World Bank (2008), "Biofuels: The Promise and the Risks" *World Development Report 2008*, pp.70-71.
2. Renewable Fuels Association, Ethanol Industry Statistics, *http://www.ethanolrfa.org/industry/statistics/#E* (accessed March 2008).
3. Pimentel, David (2003), "Ethanol Fuels: Energy Balance, Economics, and Environmental Impacts are Negative" *Natural Resources Research, Vol. 12, No. 2,* June, pp. 127-134.
4. Junginger, Martin *et al* (2006), *Sustainability of Brazilian bio-ethanol, Report NWS-E-2006-110, ISBN 90-8672-012-9*, study commissioned by SenterNovem, The Netherlands Agency for Sustainable Development and Innovation, 135 pp.
5. Kyle, Benjamin G. (1981), "Low Energy Process of Producing Gasoline-Ethanol Mixtures", *United States Patent No. 4,297,172.*
6. Rahman, M. A. *et al* (2007), "Extraction of Gasohol Grade Ethanol from Aqueous Solution using Gasoline as Solvent", *Bangladesh J. Sci. Ind. Res.* 42(3), pp. 287-298.
7. Jenkins, Renaldo V. (1999), "Separating Ethanol from Water via Differential Miscibility", *NASA Tech Briefs, http://www.techbriefs.com/content/view/1830/32/* (accessed March 2008).
8. Offeman, Richard D. et al (2006), "Solvent Extraction of Ethanol from Aqueous Solutions Using Biobased Oils, Alcohols, and Esters", *JAOCS, Vol. 83, No. 2,* pp. 153-157.
9. Boudreau, Tracy M. and Gordon A. Hill (2006), "Improved Ethanol-Water Separation Using Fatty Acids", *Process Biochemistry 41,* pp. 980-983.
10. Gil, I. D. *et al* (2008), "Separation of Ethanol and Water by Extractive Distillation with Salt and Solvent as Entrainer: Process Simulation", *Brazilian Journal of Chemical Engineering, Vol. 25, No. 01,* January-March, pp. 207 – 215.

Shortcut Methods for Scale-up of Separation Processes

Joseph B. Powell, Ph.D.
Chief Scientist – Chemical Engineering
Shell Global Solutions U.S. Inc.
Westhollow Technology Center, Houston, TX

Abstract

Shortcut methods are presented for use in the screening or scouting phase of process development, in design of the separation processes of adsorption, ion exchange, liquid extraction, and distillation. The shortcut methods may in some cases obtain all the information required for commercial process design, depending upon sensitivity of process economics to separation performance.

Overview

Much attention has been given to use of multi- or high-throughput methods for examination of reaction systems, to identify optimal traditional or biological catalysts, solvents, or reaction conditions for product synthesis. Less attention has been given to separations, though piloting of separations can often drive the scale and cost of process development pilot plants and demonstration units. The following presents a short summary of shortcut methods in design of separation processes, with emphasis on the concept that unless a great deal of time and money is spent in optimization of pilot equipment and operations, the resulting tests are imperfect representations of a commercial process. The goal of laboratory experimentation should therefore be to obtain accurate equilibrium or rate process information, to assess the separation factor achievable for the target separation. Impacts of heat and mass transfer rates on the ultimate commercial design can generally be treated by available engineering correlation, and or use of numerical models. It is typically not necessary to design and operate a large-scale pilot to re-prove these correlations, in order to succeed at scale-up.

Adsorption and Ion Exchange

Successful design of commercial processes of greater than 100 MM lb/yr annular capacity have been conducted via experiments in 10-ml glass vials augmented by small-scale laboratory column demonstration tests, with no intermediate scale piloting. Figure 1 presents the rotating rack, which represents a simple, multi-throughput apparatus for design of liquid phase adsorption and ion exchange systems. The rack may be placed in a heated oven (N_2 purge on vacuum drying oven if flammable solvents are used). An adsorption isotherm may be obtained by preparing a liquid mixture containing the solute component(s) "i" to be removed, and varying the amount of ion exchange resin or adsorbent solid charged to each vial. Crushed solids are often used, as the intent is to measure the equilibrium adsorption isotherm.

Many real systems are readily fitted to the Freundlich isotherm:

(1) $$q_i = m \bullet c_i^{\beta}$$

where q_i is the amount adsorbed per unit mass of solid, c_i is the concentration in liquid expressed as mass fraction, and isotherm β is the key parameter determining strength of adsorption . The equilibrium breakthrough capacity (T), defined in bed weights of feed which can be processed, is given by:

(2) $$T = \frac{q_i\{c_i^o\}}{c_i^o} = m \bullet q_i^{\beta-1}$$

Earlier breakthrough can occur as a result of finite rates of kinetic or transport processes. Vial tests were used to determine isotherms for two types of ion exchange resin systems examined for removal of divalent cation (Figure 2). Strong acid resin exhibited concave downward curvature indicative of strong adsorption, whereas weak acid resin at unfavorable pH = 4.4 curvature was concave upwards.

Breakthrough curves for the strong acid resin case were estimated from constant pattern solution afforded by (Vermeulen 1973), given that for liquid phase sorption with most bed designs, particle phase diffusion dominates. The solution for a linear driving force was assumed, with the number of transfer units N_p given as:

(3) $$N_p = \psi_p k_p a_p / WHSV$$

where $\psi_p \cong 1$ is a valid approximation given the strong sorption. The particle phase mass transfer coefficient is given by (Vermeulen 1973):

(4) $$k_p a_p = \frac{60 D_p}{d_p^2}$$

were d_p is the particle diameter, and the particle phase diffusivity D_p ranges from 10^{-1} to 10^{-4} of the diffusivity in the free liquid. Taking the most conservative assumption, the calculated constant pattern breakthrough was essentially a shock front corresponding to the simple mass balance of equation (2). Strong adsorption overwhelms rate processes impacting the breakthrough curve, for this system. A simple column test was conducted in a glass laboratory chromatography column (1-inch OD by 6-inch packed section of resin), to confirm the prediction. The process was taken directly to commercial scale (120 gpm flow), which validated the scale-up premise. In this case, ultimate performance was dominated more so by robustness of distributor design, and the amount of acid and contact time used for resin regeneration, as opposed to transport or rate processes associated with the ion exchange bed linear design. Additional piloting or modeling of the latter was not warranted.

The weak acid resin was also implemented commercially, despite the unfavorable (concave upwards) isotherm indicated in Figure 2. It was recognized that this isotherm could be approximated by a linear fit in the low concentration region corresponding to the trace component removal desired. Multiple mass transfer and rate processes can be considered for the case of linear isotherm and driving force, allowing analytical solution via the Klinkenberg approximation (Vermeulen 1973), valid for all mass transfer phenomenon impacting the number of transfer units N. Axial dispersion, exterior particle, and particle diffusion transfer units were separately calculated to determine limiting N_{min}. The result was validated again using the small laboratory column test, this time with multiple tests at varying fowrate to assess impact on breakthrough performance profiles. The design was directly implemented commercially, with a somewhat larger engineering safety factor (25%) to allow for lack of robust equilibrium adsorption, where effects of axial dispersion, exterior-particle, intra particle pore or particle diffusion, possible slow ion exchange kinetics, macroscopic channeling can all contribute to early breakthrough upon scale-up. Small variations in feed pH, and variability in regeneration again dominated the commercial performance.

These examples demonstrate use of small-scale vial tests to determine equilibrium, with engineering correlation used for commercial design, validated by small scale off-line column tests, as an alternative to large scale validation of bed performance in a continuous integrated pilot.

Figure 1: Rotating rack for design of ion exchange, adsorption, liquid extraction separations. Vial sizes 5 – 20 ml.

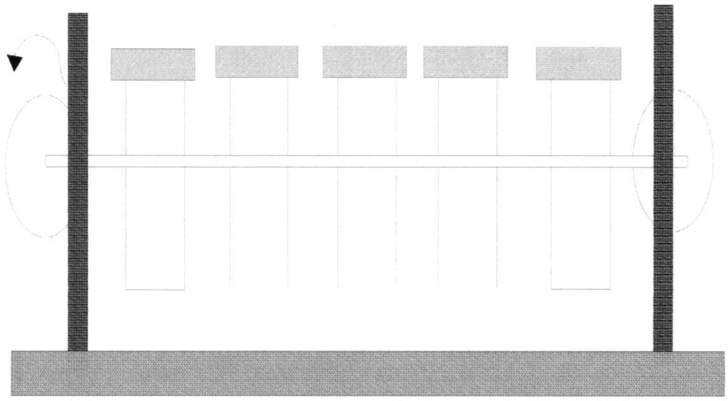

Where further diagnosis is warranted, potential slow ion exchange kinetics or transport resistances in fouled resin can be examined via use of whole bead or particles in the tumbling vials, using experiments at the same solid-liquid ratio, where a vial is removed over timed intervals from 1 – 24 hours to assess rate of uptake. Alternately, stirred V-bottom vials with resin suspended above the stir bar to avoid particle attrition, may be employed, with periodic sampling. The smaller scale batch tests can thus be adapted to study rate processes as needed, in a well-defined system.

Figure 2: Adsorption isotherms for divalent cation removal by strong and weak acid ion exchange resins.

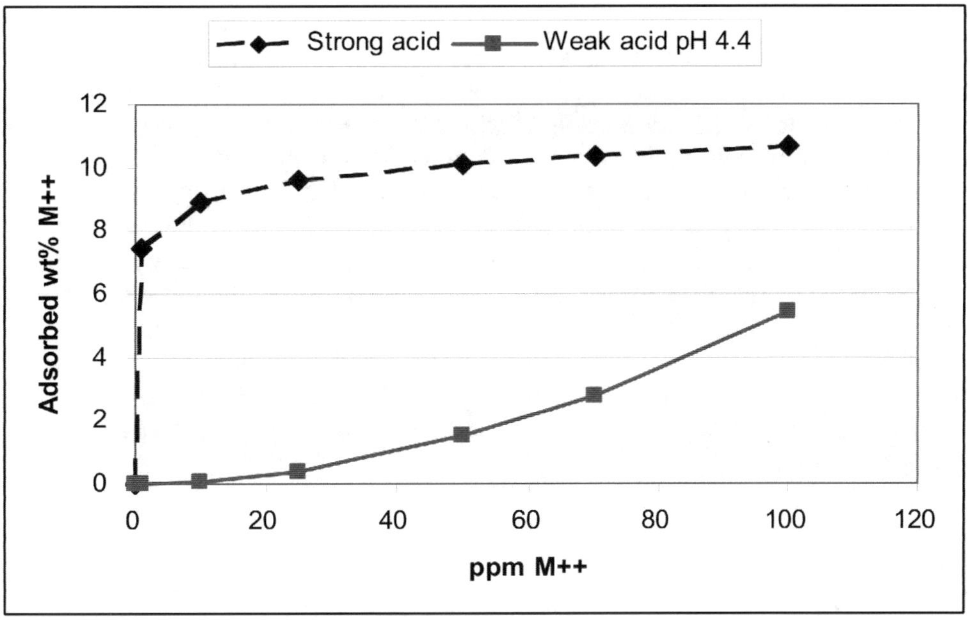

Liquid extraction

Liquid extraction design has been accomplished via experiments in the same rotating vial apparatus shown in Figure 1. Partition coefficients for the immiscible solvent pair of interest can be obtained via varying the solute concentration, and measuring upper vs. lower phase concentrations after overnight equilibration by tumbling of phases. Any concentration dependence of the L-L equilibrium can thus be evaluated. Once equilibrium has been assessed, the Kremser equation (linear dependence for equilibrium data $y = mx$) may be used for quick design estimates (Treybal 1987).

Cross current extraction may be examined batchwise for multi-component systems, entailing repeated addition of fresh solvent to treat an extract layer from the previous simulated "stage".(Treybal 1987), or extraction of fresh feed with previous cycle extract, in order to assess potential for countercurrent concentration. Use of a stirred autoclave at 25 – 300 ml volume is often sufficient for process scale-up. Concentration- dependent equilibrium data may be fitted to activity coefficient models, for use in a process simulator. Continuous countercurrent pilots can often be avoided via these approaches.

The end result is specification of the number of theoretical stages needed for a specified separation, based on equilibrium physical property measurement and model calibration. Design detail for column diameter, packed bed or tray configuration may be obtained from published correlations (Treybal 1987), or provided by a column internals vendor.

In this manner, liquid extractors for a 75 kta commercial process were designed, with no pilot testing of equipment designs, other than cross current extraction conducted in a 300-ml stirred reactor, acting as mixer-settler. Rotating vial contacting at a smaller scale (10-ml) was used to screen solvent alternatives.

The design challenge for commercial scale-up in the former example was not in the product separation, but in trace chemistry that could induce emulsion formation. Propensity for emulsion formation was assessed via a continuous integrated mini-pilot plant employing 10 ml/min of liquid feed, and a 5-hole sieve tray extractor at 2-inches in diameter. Tray efficiency was much less than commercial design under these conditions, but the extractor could be placed within the mini-pilot to test for emulsion formation with closed process recycles. In this manner, emulsion tendency was assessed independent of design information needed for scale-up of extractor separation efficiency, which was obtained from engineering correlations. This approach allowed the continuous mini-pilot to operate at 1/10 the scale of that which would have been required to simultaneously pilot both extractor efficiency and the impact of closed recycle loops. The process successfully scaled to commercial, while process development savings were in the millions of dollars.

Distillation

Distillation is another example where new separations can often be designed from batch experiments, without costly continuous large-scale piloting. The process development savings can be enormous, as a complex distillation requiring a large number of theoretical stages can readily set the scale of an integrated continuous pilot, including reaction sections needed to supply steady state feeds. Pilot distillations are typically conducted at a minimum of 1-inch diameter if temperature compensated, or 3-inch or more with higher throughputs if it is desirable to operate adiabatically to cope with lack of *a priori* compositional information needed to stipulate a temperature profile. These scales can drive pilot costs in excess of $10's of millions. Simply determining relative volatility via the method described below can often be done in a batch test at one-liter scale or less, in an existing batch still (< $25 K).

Key to this approach is use of markers of known relative volatility. Use of an Oldershaw column is the benchmark method for laboratory scale-up (Fair 2008). Markers whose activity coefficients can be calculated serve to further validate assumptions. While an Oldershaw column provides a cross check against the number of stages (50 – 60% efficiency per tray for normal organic distillations), use of higher efficiency protruded metal packing allows a potentially larger number of stages for assessment of difficult separations. The batch column can approximate operation at infinite reflux, to obtain the minimum number of stages for separation.

Per Fenske's equation (Sundaram 1993) at infinite reflux, one obtains the minimum number of stages needed for the observed separation, wherein:

(5) $$N_{\min} = \frac{\log\left[\dfrac{x_{1D} \bullet x_{2B}}{x_{2D} \bullet x_{1B}}\right]}{\log(\alpha_{12})}$$

Given α_{ij} for the known marker compounds, N_{min} can be calculated from measured overheads and bottoms compositions. With N_{min} known, α_{ij} for the unknown components can be evaluated. The approach works as long as the unknown components are not present in large concentration impacting VLE for other components, such that α_{ij} are can be calculated *a priori* for the marker components, allowing N_{min} to be assessed from the experiment. This is often the case in new process development, where separation of trace impurities is the dominant design challenge for scale-up.

The approach was successfully implemented for commercial scale-up, using a 20-foot tall by 3-inch diameter vacuum jacketed adiabatic column, operated batchwise at infinite reflux with several weeks of cumulative product from a small pilot plant. Resulting relative volatility data was used for commercial scale-up, with no continuous piloting of distillation. This reduced the size and cost of piloting approximately ten-fold.

Discussion

Pilot and demonstration plants are quite useful where development of new unit operations or testing of new process chemistry is required. Extensive correlations exist for many unit operations including separations, however. It may not be necessary to fully pilot the exact separation required for commercial scale-up. Often, smaller scale diagnoses of process chemistry can be used, together with engineering correlations, to adapt new chemistry to a known unit operation.

References:

Fair, J. R., and A. F. Seibert. 2008. Using an Oldershaw Column for Scale-up. *Chemical Engineering Progress* 104 (10):27-29.

Sundaram, S. and L. B. Evans. 1993. Shortcut procedure for simulating batch distillation operations. *Ind. Eng. Chem. Res., 1993, 32 (3),* 32 (3):511-518.

Treybal, R. E. 1987. *Mass Transfer Operations*. NY: McGraw-Hill.

Vermeulen, T., and G. Klein, N. K. Hister. 1973. Adsorption and Ion Exchange (Section 16). In *Chemical Engineer's Handbook*, edited by R. H. Perry, and Chilton, C. H. NY: McGraw-Hill.

Scale-up Technology with FinTec® Reactors

Zhijun Jia (1), Mark Wood (2), and Larry Stryker (1). Chart Energy & Chemicals, LaCrosse, WI (1) and Wolverhampton, UK (2)

Expended Abstract

Scale-up of chemical processes with highly exo- or endo-thermic reactions can be problematic. Devices used in the laboratory for process development mimic the final reactor choice, for example stirred vessels for batch processes or tubular reactors for continuous processes, with scale-up effectively increasing the dimensions of the equipment to achieve the required productivity levels. Studies indicate that the main difference between lab scale and commercial or industrial scale operation is the varying local conditions within the reactor. This variation significantly affects the mass and heat transfer and consequently the reaction products, especially for highly exo- or endothermic reactions. Achieving comparable local conditions has been a big challenge for engineers when attempting to scale up from the lab to the commercial or industrial scale.

As laboratory devices are relatively small, high levels of mixing energy input and heat transfer surface area per unit volume can be achieved. Uniform mixing or mass transfer can be obtained without any additional mixing devices and heat transfer can be balanced with the reaction heat production rate since the heat transfer path is short and the surface area for heat transfer is relatively large. On scale-up, mixing energy levels and heat transfer surface area per unit volume decrease, with longer heat transfer paths from the reaction point to the heat transfer surface. It is these changes in mass transfer and heat transfer scales that causes many of the performance reductions seen between laboratory and full scale operation. It is therefore desirable to use chemical reactors where a full scale reactor can provide the same local conditions as the laboratory scale reactor. This can be achieved through the use of compact heat exchange reactors where the channel dimensions do not change between different scales of operating. Instead, the philosophy is to have more channels of the same dimension (scaling out) rather than having a similar number of channels of larger dimensions.

Chart Energy & Chemicals, Inc (Chart), a manufacturer of compact brazed aluminium plate-fin heat exchangers for over 60 years, has been developing compact heat exchange reactors (CHER) for decades. These can be built at scales suitable for laboratory, pilot and commercial scale operation, either using ShimTec® or FinTec® technology. ShimTec is a shim-based diffusion bonded heat exchanger and reactor, and FinTec is built in plate-fin configurations with brazing technology. CHER reactors consist of separate reaction and cooling/heating streams of multiple mini-channels which are identical in geometry from lab unit through to commercial scale. Therefore the challenge in reactor scaling-up is mitigated since the mass and heat transfer conditions are maintained between the different scales.

As they are based on heat exchangers, CHER reactors significantly enhance heat transfer compared to conventional reactors, allowing improved control of the reaction. These compact features also offer process intensification (PI), with benefits including:

- Improved product quality
- Reduced residence time
- Reduced reactant inventory
- Reduced energy consumption
- Reduced capital cost for plant modifications
- Inherently safe operation

CHER reactors therefore provide their greatest benefits to faster chemical reactions that are limited by heat transfer or mass transfer. As a result the reactor footprint is significantly reduced, the process efficiency is improved and the capital and operational costs are lowered. Moreover, using CHER reactor technologies allows reactions to be operated under conditions that simply would not be possible in conventional equipment.

Chart's reactor group has successfully provided dozens of custom designed units for various applications from lab to pilot scale. Figure 1 illustrates a core made of stainless steel with FinTec technology. The unit can be used for high temperature and high pressure reaction condition, such as high temperature Fisher Tropsch process, oxidation reaction, etc. Figure 2 is an aluminium core with FinTec technology. The unit can be used for lower temperature applications, such as low temperature FT process, hydrogenation, etc. The aluminium core easily scales up for commercial application, as shown by a heat exchanger in Figure 3. It is possible to incorporate catalyst into the reactor structures, including packing catalyst into the spaces provided between the fins. A FinTec reactor undergoing packing with catalyst is shown in Figure 4. .

Figure 1 Stainless Steel FinTec core

Figure 2 Pilot scale aluminum FinTec Unit

Figure 3 Commercial scale unit

Figure 4. FinTec reactor being packed with catalyst

Conclusion

Chart's FinTec technology for compact heat exchange reactors can provide unique features for process intensification and easiness for reactor scale up. It mimics the local conditions from lab unit to commercial unit so that the mass and heat transfer can be kept in same, mitigating challenges faced during process scale up.

Efficient Process Development: Non-Technical Aspects

Cawas A. Cooper, Ph. D., P. E.

AIChE Fellow and Process Development Division Chair

Cawas.Cooper@gmail.com

This presentation will cover five non-technical aspects essential for quick and timely completion of process development projects. The second presentation will address the technical aspects for completing projects flawlessly and competitively.

The first and most important item is sponsorship by a champion. This person has to provide link to business strategy and need in addition to be visible. Obviously, the resources need to be budgeted as well.

Second most important item is a clear understanding of the objective with clear definitions of goal, target, and scope. Objective should also address financial benefit. Feed back from the team into objectives is also important.

Thirdly, the right team needs to be built so that all critical functions are represented. Usually, smaller teams move faster.

Fourth item after the team is formed, it should work as a team to make it work. Team members should know one another, technically and personally. Team building events also help to “gel” teams.

The fifth critical aspect is to have a leader for the team with defined role. This person is accountable to the team members as well as stakeholders.

Examples of successful projects that used the above five aspects will be presented. This sets the stage for starting the technical phase of the project.

Efficient Process Development: Technical Aspects

Cawas A. Cooper, Ph. D., P. E.

AIChE Fellow and Process Development Division Chair

Cawas.Cooper@gmail.com

This presentation will address the technical aspects for process development projects to be flawless and competitive. The earlier presentation covered the non-technical aspects essential for quick and timely completion of projects.

For any project, it is important to minimize costs, duration, and risks during process development of new process technology and products.

When an inquiry comes, first decision that needs to be made is where the project will be done. This decision is based on capabilities of equipment and personnel.

Once the location is selected, the following technical details have to be addressed:

Process Definition
Equipment Piping and Control Modifications
Operating Procedures
Analytical Procedures
Raw Materials
Analytical Supplies
Product and Product Packaging
Waste Definition
Administrative
Pilot Experiments
Reports
Product Shipment

Lessons learned from experience over 30 years on selecting scale, location, and scale-up steps will be presented. Methods used in technology transfer will be shared with examples.

92e Biosynthesis of the Acid by a New Technology and Bioreactors

Fatemeh Abniki, Process Engineering Dept, National Petrochemical Company(ghadir group), No.38, Street 6, Dolat Abad, Shahrake Vahdat, Tehran, Iran, Ehsan Bakhshi, R&D Center of PIDMCO , National Petrochemical Company, Tehran, Iran, and Majid Mosalla, Material Science Dept., Shiraz University, Shiraz, Iran.

The usage of a main acidifier and preservative, along with up to date tendencies to increase it consumptive demand in the world, dictates current concern in developing high-rate biological processes and modern chip technology with bioreactors. In this research we used citric acid as main acidifier to production. The choice of technology and bioreactors for biosynthesis of citric acid, as well as the efficiency of the biotechnological system as a whole ,is indicated by the kinetic relation ship for the cultivation of microorganisms and the citric acid production.

Using an adequate mathematical model for biosynthesis kinetics ,one can solve below problems :

1)current concern such as the optimal control of the process

2)the choice of rational process conditions(substrate, replacement, aeration rate, biosynthesis duration, etc)

3) technology and bioreactor implementation of the process

In order to analyze of a complex biotechnological system, it is advisable to formulate its total effectiveness index as an addition function of partial in indices of the subsystems that make up it. We consider a mathematical function with 3 parameters where X are state variables, U are control variables and Y are output parameters.

The analysis of the biotechnological system for citric acid producing allowed us to identify two basic subsystems :

1- The biosynthesis subsystem, where the initial carbon-containing substrate at the stage of microbiological synthesis is transformed into the acid which is excreted to the culture broth by the producing strains during their metabolism ,and membrane module.

2- The further stage of isolation as crystallization and drying.

The optimization criteria Φ which arises out of the parameters, has been defined as for the optimal calculation of the chip bioreactor in accordance with the special algorithm we have made. It has been taking into account the biosynthesis time ,concentration of biomass, citric acid and substrate, and parameters of membrane.

Finally, the approach which is based on the mathematical models of kinetics and membrane module and the general algorithm for optimization, allowed us to design plants for the biosynthesis of citric acid, to predict the optimal process parameters and to optimize the performance of operating biotechnological plants.

Web Page: www.nipc.net

92f Sensible Heat Storage by Supporting Phase Change Material, PCM

Fatemeh Abniki, Process Engineering Dept, National Petrochemical Company(ghadir group), No.38, Street 6, Dolat Abad, Shahrake Vahdat, Tehran, Iran and Ehsan Bakhshi, R&D Center of PIDMCO , National Petrochemical Company, Tehran, Iran.

The sensible heat storage, larger amounts of energy per unit volume can be stored by utilizing the latent heat of a phase change . In this case, during the daytime hours the incident sunlight shining onto the building is stored as the salt melts. Then at nighttime ,as the salt solidified, the released energy can heat the building .Recently, latent heat " cold storage ' has been proposed as a means of leveling air conditioning loads in hot climates. One scheme, which is being commercialized, uses latent heat to store waste heat from an automobile engine so that it may be used to preheat the engine in cold climates following being off for more than several hours. It is evident that in a typical packed bed regenerator, far greater amounts of energy can be stored if the packing wee replaced with material which changes phase and stores energy as latent heat.

In this case , the heat storage medium is some means of containing or supporting the phase change materials, PCM, must be provided since the material can flow when it is in the liquid state .Models for the phase change regenerator, PCR ,are complex , involving the solution of the classical Stefan problem for phase change heat storage on the scale of the PCM, coupled with the heat balance equation for the scale of the PCR packed bed. The model presented accounts for convection, heat accumulation in the fluid ,heat losses through the wall of the PCR and energy storage by PCM. A PCM consisting of water in spherical support polypropylene (P.P) is used in a laboratory scale PCR to verify the model.

Experiments with heated or cooled air passing through the PCR are described. The measured outlet temperature compares qualitatively with the model predictions. With the importance of heat recovery technology to reduce energy consumption, there is to develop an understanding for the PCR operation.

We have presented a computational scheme for the prediction of temperature profiles within the bed and the outlet fluid temperature .

Parametric studies have verified that for all practical purpose , the ideal PCR is approached for: (3Bi)/St<5*10; Pe > 40 ; St > 50

The computational model which was developed is comprehensive and robust. It is capable of assessing the effect of operating and design parameters and performance of commercial-size PCRs.

Comparisons made between the outlet fluid temperature which was measured experimentally to that predicted by the model indicate that, in the small insulated bench-scale apparatus, there is a signification amount of heat loss. As much as 50% of the energy stored in the experimental unit can be lost to the surrounding via the end plates.

AUTHOR INDEX